John Greaves

A Treatise on Elementary Statics

Second Edition

John Greaves

A Treatise on Elementary Statics
Second Edition

ISBN/EAN: 9783337277635

Printed in Europe, USA, Canada, Australia, Japan

Cover: Foto ©berggeist007 / pixelio.de

More available books at **www.hansebooks.com**

A TREATISE

ON

ELEMENTARY STATICS.

ON

ELEMENTARY STATICS

BY

JOHN GREAVES, M.A.,

FELLOW AND MATHEMATICAL LECTURER OF CHRIST'S COLLEGE, CAMBRIDGE;
FORMERLY ASSISTANT MASTER AT BEDFORD GRAMMAR SCHOOL.

SECOND EDITION.

London:
MACMILLAN AND CO.
AND NEW YORK
1888

[*All Rights reserved.*]

First Edition published in 1886.
Second Edition, 1888.

PREFACE.

THERE has existed for some time past a general feeling that the Laws of Motion form the only satisfactory basis on which the science of Statics can be built. So far as I know, all the text-books in use in Cambridge except Prof. Minchin's, treat the subject from quite a different point of view. In this text-book I have endeavoured to supply the wants of students, who are not sufficiently advanced in Pure Mathematics, to read with advantage Professor Minchin's treatise on Analytical Statics.

Deducing from the Newtonian definition of force and the parallelogram of velocities, the *parallelogram of forces*, I obtain the *necessary* conditions of equilibrium for *any* material system by means of the third law, without assuming the *transmissibility of force*, or supposing the system to become rigid. From these and certain geometrical considerations follow the *sufficient* conditions of equilibrium of a rigid body. This involves the introduction of the conception of the moment of a force about a line, and certain geometrical propositions, which may be regarded as somewhat difficult for a beginner: I am in hopes that these difficulties will not be found insuperable, as it seems to me that there is a distinct gain in clearness and simplicity by this mode of treatment of the subject. The Appendix on indefinitely small quantities has been added to enable the student, who is unacquainted with Newton's *Lemmas* and the *Differential Calculus,* to follow the methods used in the chapters on the Centre of Mass and Virtual Work.

For the sake of students beginning the subject, easy numerical examples on the preceding propositions have been embodied in the text. The articles, marked with an asterisk, may be reserved for a second reading of the subject. Explanations and illustrations are printed in smaller type than the articles relating to general principles. With the view of making the diagrams more intelligible, the bounding lines of physical surfaces are drawn thicker than lines representing forces, and lines drawn merely to obtain a geometrical solution of the problem are dotted.

I have referred continually to Thomson and Tait's *Natural Philosophy,* and have also consulted Jellett's *Theory of Friction.* Several of the Illustrative Examples are taken from Dr Wolstenholme's Collection.

I am much indebted to Mr E. W. Hobson, M.A., Fellow of Christ's College, for many valuable suggestions, and also to Mr J. B. Holt, B.A., Scholar of Christ's College, for his kind criticisms and assistance.

JOHN GREAVES.

CHRIST'S COLLEGE, CAMBRIDGE,
    April, 1886.

PREFACE TO THE SECOND EDITION.

A FEW articles, which seemed wanting in clearness, have been re-written; and I have introduced Coulomb's empirical formula for taking into account the rigidity of ropes used in machines, as well as a description of Weston's Differential Pulley. Many errors in Examples have been corrected, and it is hoped that very few now remain.

April, 1888.

CONTENTS.

CHAPTER I.

ON THE STATICS OF A SINGLE PARTICLE.

	PAGE
Parallelogram of velocities	5
Parallelogram of forces	11
Triangle of forces	14
Polygon of forces	15
Lami's theorem	16
Magnitude of resultant of two forces	17
Resolution of forces	22
Conditions of equilibrium	31
Theorem relating to moments about a point	34
Conditions of equilibrium	34
Theorem relating to moments about a line	36
Illustrative Examples	40
Examples on Chap. I.	49

CHAPTER II.

ON THE STATICS OF SYSTEMS OF PARTICLES.

Necessary conditions of equilibrium	54
Sufficient conditions of equilibrium for coplanar forces	59
Transmissibility of force	60
Sufficient conditions of equilibrium for any system of forces	62
Resultant of two parallel forces	67
Theorem relating to three forces in equilibrium	68
Equivalence of couples	71

	PAGE
Couples follow parallelogram law	74
Poinsot's central axis	77
Illustrative Examples	80
Examples on Chap. II.	88

CHAPTER III.

ON THE STATICS OF CONSTRAINED BODIES, &c.

Condition of equilibrium for coplanar forces, one point fixed	97
Condition of equilibrium for any forces, two points fixed	98
Conditions of equilibrium for any forces, one point fixed	99
Bending moment	101
Funicular polygon	102
Tension of string stretched over smooth surface	104
Illustrative Examples	105
Examples on Chap. III.	113

CHAPTER IV.

ON CENTRES OF MASS.

Position of centre of parallel forces :	122
Centre of gravity	126
Theorem relating to centres of mass	127
Centre of mass of a triangle	129
Centre of mass of a triangular pyramid	134
Centre of mass of arc of circle	138
Centre of mass of sector of circle	140
Centre of mass of zone of sphere	141
Centre of mass of segment of parabola	145
Centre of mass of rod of varying density	146
Condition of equilibrium of a body placed on a horizontal plane	148
Illustrative Examples	151
Examples on Chap. IV.	155

CHAPTER V.

On Friction.

	PAGE
Laws of friction	164
Empirical laws of limiting friction	165
Tension of string stretched over rough surface	169
Nature of initial motion	172
Illustrative Examples	173
Examples on Chap. V.	185

CHAPTER VI.

On Virtual Work.

Principle for single particle	198
Principle for rigid body	199
Cases of forces which do no work	202
Converse of principle for single particle	205
Converse of principle for rigid body	206
Work done during initial motion necessarily positive	207
Positions of stability and instability	208
Stability of body resting on rough surface	210
General theorem relating to potential energy	215
Illustrative Examples	216
Examples on Chap. VI.	225

CHAPTER VII.

On Machines.

Lever	231
Pressure on the fulcrum	232
Efficiency of a rough lever	235
Wheel and axle	235
Rigidity of ropes	237
Single pulley	239
First system of pulleys	240

	PAGE
Second system of pulleys	242
Third system of pulleys	243
Inclined plane	246
Screw	249
Balance	254
Requisites of a good balance	255
Common steelyard	257
Danish steelyard	258
Roberval's balance	260
Differential wheel and axle	261
Weston's differential pulley	262
Hunter's differential screw	263
Examples on Chap. VII.	264

APPENDIX.

Guldin's theorems	270
Limit of a certain infinite series	272

STATICS.

CHAPTER I.

STATICS OF A SINGLE PARTICLE.

1. When a point is changing its position relatively to surrounding points, it is said to be *in motion* relatively to them: if it is not changing its position, it is said to be *at rest*.

If we consider not only the actual change of position, but also the time which the motion occupies, we bring in the idea of *rate of motion* or *velocity*.

2. *Def.* If a point moves over equal distances, in equal successive intervals of time, no matter how short the intervals are, the velocity of the point is said to be *uniform*. If the distances are not equal, the velocity is *varying*.

For the velocity to be uniform, it is essential that the distances be equal, even when the intervals of time are indefinitely small: for instance, we may imagine a train travelling 30 miles during each of several successive hours, yet we should not describe its motion as uniform, if the distances travelled during the different minutes were not all equal, nor yet, even though the distances travelled during the different minutes were so, provided those travelled during the different seconds were not always the same, and so on indefinitely.

3. If we wish to give any one a clear idea of the magnitude of some physical quantity, we describe it as bearing such and such a ratio to some definite arbitrarily chosen amount of that quantity, known to him. The known definite amount is called the *unit* of the physical quantity generally, while the ratio is called the *numerical measure*, or simply the *measure* of the particular amount under consideration.

If for instance, the area of a certain field be $12\frac{1}{2}$ acres, and an acre be chosen as the unit area, the ratio of the area of the field to that of the unit is $12\frac{1}{2}$, which is therefore the numerical measure of the area of the field.

We shall suppose then, that we have fixed on some particular length as the unit length, and some particular interval of time as the unit of time.

If the velocity of a point be uniform, its numerical measure is the numerical measure of the distance traversed by it during the unit of time. It may happen that the point's velocity, though uniform for a finite interval of time, is not so for the unit of time: in that case, its numerical measure is that of the space the point would traverse during the unit of time, provided it moved throughout with the same velocity as during the finite time. The velocity which we call the unit velocity, or whose numerical measure is *one*, is the velocity of a point which traverses the unit of length in the unit of time.

Def. The *mean* or *average* velocity of a point during any interval of time is the velocity with which a point, moving uniformly during that time, would describe the same distance. Its numerical measure is therefore the numerical measure of the distance described, divided by that of the time required.

Def. The velocity of a point *at any instant*, is the limit of the mean velocity of the point during an interval of time including the particular instant, when the interval is diminished indefinitely.

STATICS OF A SINGLE PARTICLE. 3

Ex. 1. Compare the velocities of two points which move uniformly, one through 5 feet in half a second, and the other through 100 yards in a minute. *Ans.* 2 : 1.

Ex. 2. A railway train travels 160 miles in 6 hours 30 minutes. What is its average velocity in feet per second? *Ans.* 36·1 nearly.

Ex. 3. One point moves uniformly twice round the circumference of a circle, while another moves uniformly along the diameter: compare their velocities. *Ans.* $2\pi : 1$.

Ex. 4. A fly-wheel is 14 feet in diameter, and is observed to go round uniformly fifteen times in a minute: find the velocity of a point in the circumference. *Ans.* 11 feet per second nearly.

Ex. 5. Supposing the earth to rotate about its axis in 23 hours 56 minutes, its equatorial diameter being 7925 miles, find the velocity of a point at the equator relative to the earth's centre, in feet per second. *Ans.* 1526 nearly.

4. Now a velocity is entirely known, if its *direction* and *magnitude* are known. But as a straight line AB can be drawn in any direction, it can be drawn so as to indicate fully the *direction* of a point's velocity, provided we shew either by an arrow-head or by the order of the letters AB, the *sense* of the velocity, i.e. whether its direction be from A to B or from B to A. As we can make the line of any length, we can make it so that its length bears the same ratio to some arbitrarily chosen length as the velocity considered bears to the unit of velocity. If this be done, and we know the *scale*, i.e. the length chosen to represent the unit velocity, the line AB will also represent the *magnitude* of the velocity considered.

Fig. 1

5. A point may be moving with several independent velocities at once: for instance, we know that the earth as a whole is describing an orbit about the sun, and that all points on the earth's surface are describing circles about the earth's axis; if then, a point be moving on the earth's surface, it has relatively to the sun, three independent velocities, viz. its velocity on the earth's surface, the velocity of the point

1—2

of the earth's surface it occupies at the particular instant, relatively to the centre of the earth, and the velocity of the earth's centre about the sun.

Def. When a point has several independent velocities, the single velocity which would alone give the point's motion is called the *resultant* of the other velocities.

Let us consider the case of a point moving in a straight line along the deck of a ship, with uniform velocity relative to the ship, which is sailing with uniform velocity in a straight line along the earth's surface. It is required to find the point's motion relative to the earth's surface, i.e. given its position at one instant, it is required to find its position at the end of a given time. Now since the point's motion on the ship's deck is entirely independent of the ship's motion, if we suppose the point fixed to the deck during the time considered, so that its motion is that of the ship, then the ship to remain stationary while the point moves for an equal time along the deck with its velocity relative to the ship, the final position of the point will be the same as if the two motions had taken place simultaneously, as they really do.

The above illustration exemplifies a general axiomatic principle, which may be stated thus: *if during a certain time a point has several independent motions, its actual position at the end of any portion of that time may be found by imagining that all the motions take place separately during a number of successive periods of time equal to the one considered, instead of supposing that all the motions take place simultaneously, which is what really takes place.* Of course the imaginary motion only gives the same initial and final positions of the point as the real one, and not in general intermediate ones, although by taking the periods of time very small, but very large in number, the imaginary motion which gives us the real position of the point at the end of each of them, will give us an infinite number of points on the point's actual path. The motions referred to above are not of necessity due to uniform velocities.

The Parallelogram of Velocities.

6. *If the two independent velocities of a point be represented in magnitude, direction and sense by two straight lines drawn from, (or to) a point, and a parallelogram be constructed on them as adjacent sides, the resultant velocity is represented in magnitude, direction and sense by the diagonal drawn from (or to) the point of intersection of these sides.*

Let the lines OA, OB represent in magnitude, sense and direction the velocities u, v of the point: complete the parallelogram $OACB$, and join OC; then OC shall represent the resultant velocity. If O be taken as the

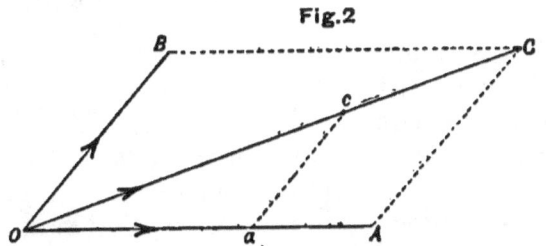

Fig.2

initial position of the point, its position at the end of a time t can be found by supposing that it first moves with the velocity u for a time t, and then with the velocity v for the same time. If it moves with the velocity u alone, it will at the end of a time t be at a in the line OA, where $Oa = ut$; if now it moves with the velocity v alone for a time t, it will arrive at c, where ac is parallel to OB, and $ac = vt$. c then is the position of the point at the end of a time t, when the motions take place simultaneously.

But because $ac : Oa = AC : OA$, c is in OC, i.e. OC represents the *direction* of the resultant velocity. Also the magnitude of the resultant velocity is to the velocity OA as Oc is to Oa, i.e. as $OC : OA$: hence OC represents the *magnitude* of the resultant velocity. The *sense* of the resultant velocity is clearly OC.

The above proposition holds at any instant, even though the independent velocities be varying velocities: for it is only necessary to suppose that the time t is ultimately indefinitely small, and the above proof holds.

Ex. 1. If a boat can steam 9 miles an hour up stream, and 13 miles an hour down stream: find the velocity of the stream.

Ans. 2 miles per hour.

Ex. 2. Velocities of 4 feet and 16 feet per second in directions at right angles to each other are simultaneously communicated to a body: determine the resultant velocity. *Ans.* 16·49 feet per second.

Ex. 3. A ship whose head points N.E. is steaming at the rate of 12 knots an hour in a current which flows S.E. at the rate of 5 knots an hour, find the velocity of the ship relative to the sea bottom.

Ans. 13 knots an hour.

7. All the objects around us that we can see and touch, and even invisible substances, such as air, are *material* bodies or composed of *matter*. The various properties of matter, such as hardness, density, &c., can be investigated, but no definition of matter can be given which would give any idea of it to a being that had had no experience of it.

Any limited portion of matter is called a *Material Body* or simply a *Body*. When we consider a body whose dimensions are so small that we are only concerned with its motion as a whole, and not with any rotational motion it may have, we describe it as a *material particle*, or simply a *particle*.

The term *Mass* is synonymous with the phrase *Quantity of Matter*, so that the mass of a body means the quantity of matter in it.

If two bodies are composed of the same substance under the same conditions we can compare their *masses* by comparing their *volumes*. Thus a quart of water contains twice as much matter as a pint. Liquids generally are sold by volume. In the case of solids it is often difficult to determine their volumes and when substances are of different kinds we cannot assume that their masses are proportional to their volumes. We shall learn in Art. 9 how to find the ratio one mass bears to another.

The *Momentum* of a body is measured by the product of its mass into its velocity. Its direction is the same as that of the velocity.

8. *Statics* is the science which treats of the equilibrium of bodies under the action of *Forces*. A definition of the term *Force* is supplied by Newton's 1st Law, which asserts that '*Every body remains in a state of rest or of uniform motion in a straight line, except in so far as it may be compelled by impressed forces to change that state*'.

Force, then, is that which alters or tends to alter the state of rest or of uniform motion in a straight line, of a body. It is not necessary to suppose that the state of rest or of uniform motion is *actually* altered by a force, because other forces may be in action which counteract the effect of the first. If we observe a body moving in any way other than uniformly in a straight line, we infer that it is acted on by force: e.g. when we find that the planets move in nearly elliptic orbits, we know that each is under the action of some force: similarly when we see that a falling body moves with gradually increasing velocity, or that another is stopped, we know that a force has acted on each of them. If a force acts for a time on a body, producing a change in the body's velocity, it is clear that if it continues to act, it will tend to produce a still further change.

We are all of us familiar with some instances of the manifestation of force. For instance, we may set a body in motion or stop it by pushing it, either directly with the hand, or by means of a rod, or we may pull it by a string attached to it: we may also expose it to the action of the wind or to the pressure of steam. In all these cases the force is exerted by tangible means, but force is often manifested without any tangible means, as in the case of *gravity*, the name given to the force which causes any body near the earth to move towards it, and the planets to revolve in their orbits about the sun; also in the case of the force which causes small pieces of iron to move towards a magnet held near them. A force of this kind is called an *attraction*.

9. The next question that presents itself is 'How is force measured?' or 'When may *this* force be said to bear such and such a ratio to *that* force?' We know by experience that it requires a greater effort on our part to impart a given velocity to a large amount of any substance than to a small amount, but what determines the exact ratio that exists between the two forces?

Our own sensations do not give us an accurate scale by which the forces may be measured. The answer is contained in Newton's 2nd Law, which asserts that '*Change of motion is proportional to the impressed force and takes place in the direction of that force*'.

By *Change of Motion* is meant *Change of Momentum in some fixed time*, the unit of time for instance. *The Change of Motion* is therefore the product of the mass acted on into the change of velocity produced in a given time.

First, let us suppose that a number of different forces act on the same particle during equal intervals of time, so that the only variations in the different cases are the differences in the forces and in the changes of velocity produced as the mass in each case is the same. Hence forces are *equal* if, when they act for equal times on the same particle, they produce equal changes of velocity, and the ratio between the magnitudes of two forces is the ratio between the respective changes of velocity they produce in the same particle, after acting for equal times.

The *direction* of a force is clearly defined by the latter part of the law as the direction of the change of velocity produced by the force.

It is of course to be understood that the change of velocity meant is not necessarily the increase or decrease of the particle's velocity, but that velocity which, compounded with the particle's initial velocity, will give the final velocity.

Next, let us suppose that a number of forces act one on each of a number of particles for the same time, and produce the same changes in their velocities. Since the change of velocity produced in each case is the same, the forces are proportional to the masses on which they act. Hence masses are *equal* when equal forces produce in them in equal times, the same change of velocity. Also the ratio between two masses is the ratio between the forces which produce in them the same change of velocity in the same time.

10. In both theoretical calculations and in actual practice we must fix on some standard force which is to be the unit. In theoretical calculations we take as our unit the *dyne*, which is the force required to generate in one second a velocity of 1 centimetre per second in a mass equal to that of a cubic centimetre of distilled water at $4°$ C. This is called the *absolute unit* and the advantage in its use is, that all the terms involved in its definition are the same at all points of the earth's surface and indeed everywhere.

It is found that if bodies be allowed to fall towards the earth in a vacuum, so that the air does not resist their motion, the velocities with which they fall are increased every second by an amount always the same at the same point of the earth's surface, and nearly so all over it. The force which produces this change of motion in a body is called its *Weight*: hence the weights of different bodies are proportional to their masses, since the change of velocity produced is the same for all. It is on this account that we generally ascertain the *mass* of a body by *weighing* the body. Assuming for the present, what we shall prove hereafter (Art. 13), that when a body is at rest under the action of two forces, they are equal in magnitude and opposite in direction, we see that the force required to support a body is equal and opposite to its weight, and would, if it acted alone, produce in the body the same change of motion upwards that its weight does downwards.

In practice in Statics, forces are generally measured in terms of the weights they would support if they acted upwards: for instance in England that force that is just sufficient to support a certain lump of metal kept at the Exchequer and called the Imperial Pound, is very often regarded as the unit force, the slight variations in this force at different places being of little consequence for practical purposes.

The velocity of a falling body is increased every second by 32 feet per second, approximately.

Ex. 1. If a body weighing 60 lbs. be moved by a constant force which generates in it in a second a velocity of 5 feet per second, find what weight the force would statically support. *Ans.* 9·3 lbs. nearly.

Ex. 2. During what time must a constant force equal to the weight of one ton act upon a train of 100 tons to generate in it a velocity of 40 miles an hour? *Ans.* 3 min. 3$\frac{1}{3}$ secs.

Ex. 3. A force which can statically support 25 lbs. acts uniformly for one minute on a mass of 400 lbs.: find the velocity acquired by the body. *Ans.* 120 feet per second.

Ex. 4. Find what velocity a force which would support a weight of n lbs. will give a mass of m lbs. in t secs. *Ans.* $32nt/m$ feet per sec.

11. *Def.* The *resultant* of a number of forces is that *single* force whose effect is the same as that of the original forces.

It frequently happens that there are several forces acting simultaneously on a body: e.g. a kite in the air is acted on by its weight, by the pressure of the wind and by the tension of the string attached to it. In the case of a particle acted on by several forces, we shall shew that there is a single force which could produce exactly the same effect as the other forces do.

The second Law of motion states that the change of motion is proportional to the impressed force and takes place in the direction of that force, so that if there are several impressed forces we infer that the actual motion will be the resultant of the several independent motions which the forces would produce if they acted separately, because the law holds for each force, and therefore these independent motions must each be produced. But this resultant change of motion might be produced by a single force, which is therefore the resultant of the original forces.

12. When a particle is in equilibrium or moving uniformly in a straight line under the action of a number of forces there is no change of motion, and therefore the resultant force must be zero; conversely, when the resultant force is zero, there is no change of motion, and the particle must be at rest or be moving uniformly in a straight line. The necessary and sufficient condition of a particle's being in equilibrium under the action of a number of forces is that their resultant be zero.

Note. Strictly speaking, if the resultant force on a particle is zero, it only shews that the particle's velocity is undergoing no change, and not that it is necessarily zero. As however in this subject we always suppose the particle initially at rest, if this condition holds, it will always remain so.

13. We have already inferred from Newton's second Law, that the direction of a force is that of the change of motion it produces, and that its magnitude is proportional to that of the change of motion: hence a force is completely defined when the magnitude and direction of the change of velocity it produces in a given time, in a particle of given mass, are given. A force may therefore be represented completely by the straight line that represents this change in velocity. We are now in a position to prove the following most important proposition, known as the *Parallelogram of Forces*.

Prop. *If two straight lines be drawn from (or to) a point representing in magnitude, direction and sense, forces acting on a particle, and a parallelogram be constructed having these two lines as adjacent sides, the diagonal drawn from (or to) the point mentioned will represent the resultant completely.*

Let OA, OB be two straight lines representing the magnitude, direction and sense of two forces acting on a particle. Complete the parallelogram *OACB*, having *OA, OB* for two adjacent sides, and join *OC*.

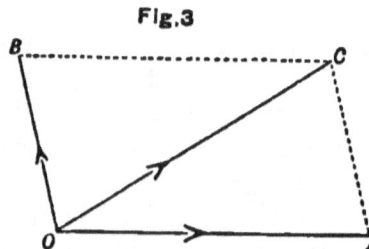

Fig.3

Since *OA, OB* represent the forces completely, they also represent the changes in velocity they would separately produce in a certain time in a particle of certain mass. By the parallelogram of velocities then, *OC* represents the resultant change of velocity they would produce in the same time in the same particle, and therefore represents the resultant of the original forces.

The following particular case of this proposition is very important. *Since the diagonal of a parallelogram always has a finite length unless the two adjacent sides are equal in length and in opposite directions, the resultant of two forces is never zero, i.e. two forces do not counterbalance one another, unless they are equal in magnitude and opposite in direction.*

Cor. If three forces not in one plane acting on a particle, be represented in every respect by three lines *OA, OB, OC* drawn from a point, and a parallelopiped be constructed on these lines as adjacent edges, the diagonal *OG* of the parallelopiped represents the resultant in every respect.

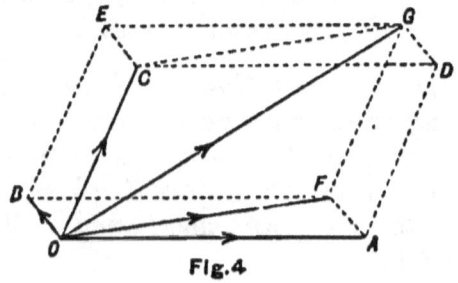

Fig.4

For *OF* is clearly the resultant of *OA* and *OB*, and since *OFGC* is a parallelogram (*OC, FG* being equal and parallel), OG is the resultant of OF and OC, i.e. of OA, OB, OC.

Ex. 1. Find the resultant of two forces of 12 lbs. and 35 lbs. respectively, which act at right angles on a particle. *Ans.* 37 lbs.

Ex. 2. If two forces acting at right angles to each other be in the proportion of 2 to $\sqrt{5}$, and their resultant be 9 lbs. find the forces.

Ans. 6 lbs., $3\sqrt{5}$ lbs.

Ex. 3. The resultant of two forces which act at right angles on a particle is 51 lbs.: one of the components is 24 lbs.: find the other.

Ans. 45 lbs.

Ex. 4. Two forces act on a particle, and their greatest and least possible resultants are 17 lbs. and 3 lbs.: find the forces.

Ans. 7 lbs., 10 lbs.

Ex. 5. Two forces acting in opposite directions to one another on a particle have a resultant of 28 lbs.: and if they acted at right angles they would have a resultant of 52 lbs.: find the forces. *Ans.* 48 lbs., 20 lbs.

Ex. 6. Two forces, one of which is three times the other, act on a particle, and are such that if 9 lbs. be added to the larger, and the smaller be doubled, the direction of the resultant is unchanged: find the forces.

Ans. 9 lbs., 3 lbs.

Ex. 7. Shew that if the angle at which two given forces are inclined to each other is increased, their resultant is diminished.

Ex. 8. If the resultant of two forces is at right angles to one of the forces, shew that it is less than the other force.

Ex. 9. If the resultant of two forces is at right angles to one force and also equal to the other divided by $\sqrt{2}$, compare the forces.

Ans. $1 : \sqrt{2}$.

Ex. 10. Two forces are represented by two chords of a circle, drawn from a point on the circumference at right angles to one another: shew that the resultant is represented by the diameter which passes through the point.

14. **Prop.** *If a particle be in equilibrium under the action of a number of forces, any one of them is equal and opposite to the resultant of the rest.*

From the definition of a resultant, all the forces but one can be replaced by their resultant without altering their effect, so that this resultant force and the remaining force maintain equilibrium, which we have seen can only be the case when they are equal and opposite.

15. The following proposition known as the *Triangle of Forces* is practically another way of stating the Parallelogram of Forces.

If three forces acting on a particle can be represented in magnitude, direction and sense by the sides of a triangle, taken in order, the forces are in equilibrium.

By the phrase '*taken in order*' is meant, that the arrowheads which indicate the directions of the forces, should all point the same way round the triangle, or that no two should both point to or from the same point.

Let ABC be a triangle whose sides AB, BC, CA, taken in order, represent in magnitude, direction and sense three forces acting on a particle—the particle shall be in equilibrium.

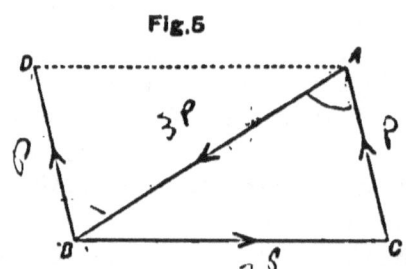

Fig. 5

Complete the parallelogram $BCAD$. Since BD is equal and parallel to CA, it will represent the force represented by CA: but the resultant of the forces represented by BC, BD is represented by BA, and is therefore counterbalanced by the force represented by AB, so that the three forces produce equilibrium.

16. Conversely, *if three forces keep a particle in equilibrium, and a triangle be drawn having its sides parallel to the directions of the forces respectively, the sides are proportional to the forces to whose directions they are respectively parallel.*

Let BC, BD (fig. 5) represent two of the forces: then, since they are in equilibrium, AB must represent the third. But CA is parallel and equal to BD, therefore the triangle ABC has its sides parallel to the three forces, and also proportional to them respectively. Any triangle then, that has its sides parallel to the three forces respectively, must have them parallel to the sides of the triangle ABC, and must therefore be equiangular to this triangle: equiangular triangles are similar ones, so that the forces are proportional to the sides, to which they are respectively parallel, of any triangle drawn in the way described.

This proposition may be extended thus: if three forces keep a particle in equilibrium, and a triangle be drawn with its sides making a constant angle measured in the same direction, with the directions of the forces respectively, the sides of the triangle are respectively proportional to the forces with whose directions they make the constant angle.

For if the triangle be turned in its own plane through an angle equal to the constant angle, but in the direction opposite to that in which the angle is measured, each of its sides becomes parallel to the direction with which it previously made the constant angle, and the proposition becomes identical with the previous one.

17. The Triangle of Forces can be easily extended to the *Polygon of Forces*, which is: *If a particle be under the action of a number of forces, which can be represented by the sides of a polygon taken in order, the particle will be in equilibrium.*

Let the sides AB, BC, CD, DE, EF, FG, GA of the polygon $ABCDEFG$, taken in order, represent a number of forces acting on a particle. Join AC, AD, AE, AF.

By the Triangle of forces, the forces represented by AB, BC can be counterbalanced by CA, therefore AC represents their resultant; similarly the resultant of AC, CD is represented by AD, that of AD, DE by AE, that of AE, EF by AF, and that of AF, FG by AG; therefore the resultant of forces represented by AB, BC, CD, DE, EF, FG is represented by AG; but forces represented by AG, GA counterbalance one another, so that the original forces are in equilibrium.

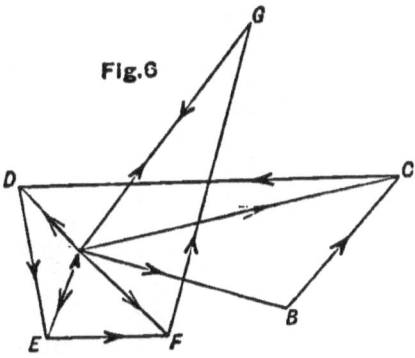

Fig. 6

16 STATICS.

Note. The forces are not necessarily in one plane.

Cor. To obtain geometrically the resultant of a number of forces acting on a particle. Draw a series of straight lines, end to end, AB, BC, CD, DE, EF, FG to represent completely the forces, whose resultant is required, then join AG, it represents completely the resultant.

The following particular case of the polygon of forces may be noticed: the resultant of a number of forces on a particle, and in the same straight line, is their algebraical sum, the forces being estimated positive in one direction and negative in the other.

The converse of the polygon of forces does not hold, because equiangular polygons are not necessarily similar.

18. The following theorem, enunciated by Lami, is the parallelogram of forces in another form.

Prop. *If three forces acting on a particle, keep it in equilibrium, each is proportional to the sine of the angle between the other two.*

Let OA, OB, OC represent three forces P, Q, R which, acting on a particle, keep it in equilibrium.

With OA, OB as adjacent sides complete the parallelogram $OADB$: join OD.

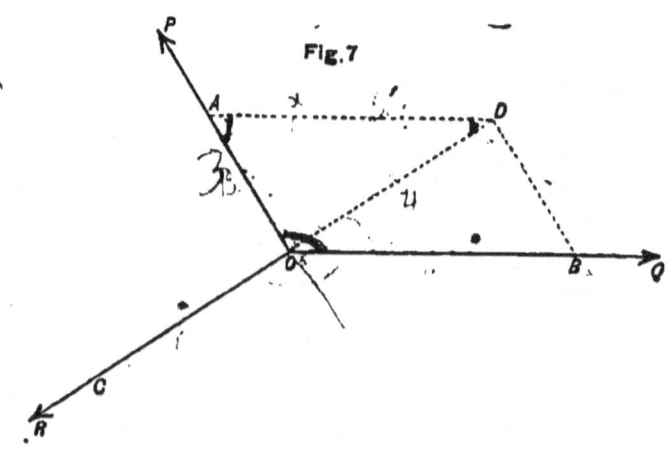

Fig. 7

STATICS OF A SINGLE PARTICLE. 17

OD, OC must be equal and opposite, since OD represents the resultant of P and Q.

$$P : Q : R = OA : OB : OC = OA : AD : OD$$
$$= \sin ODA : \sin AOD : \sin OAD$$
$$= \sin DOB : \sin AOD : \sin AOB$$
$$= \sin BOC : \sin AOC : \sin AOB$$
$$= \sin (Q, R) : \sin (P, R) : \sin (P, Q).$$

19. The magnitude of the resultant R, of two forces P and Q, which act on a particle, and whose directions make an angle θ with one another, may be easily found.

Let OA, OB represent the forces P, Q respectively. Complete the parallelogram $OBCA$, and join OC: the latter represents R.

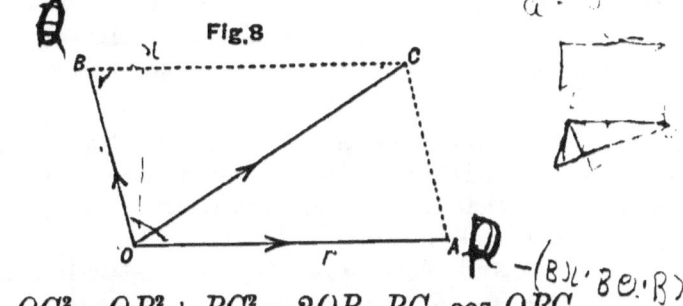

Fig.8

But $OC^2 = OB^2 + BC^2 - 2OB \cdot BC \cdot \cos OBC$,
$BC = AO$, and $OBC = 180° - AOB$
$= 180° - \theta$,
$\therefore R^2 = P^2 + Q^2 + 2PQ \cos \theta.$

Ex. 1. If forces of 3 lbs. and 4 lbs. have a resultant of 5 lbs., at what angle do they act? *Ans.* 90°.

Ex. 2. If one of two forces acting on a particle is 5 lbs., and the resultant is also 5 lbs., and at right angles to the known force, find the magnitude and direction of the other force.

Ans. $5\sqrt{2}$ lbs., making an angle of 135° with the other force.

Ex. 3. At what angle must forces P and $2P$ act on a particle in order that their resultant may be at right angles to one of them? *Ans.* 120°.

G.

Ex. 4. If three forces, whose magnitudes are expressed by the numbers 3, 6, 9, act on a particle, and keep it at rest, shew that they must all act in the same straight line.

Ex. 5. Shew that three forces cannot maintain a particle in equilibrium if one of them be greater than the sum of the other two.

Ex. 6. Find the magnitude of the resultant of (i) 3 lbs. and 4 lbs. at $45°$ to one another; (ii) 2 lbs. and 3 lbs. at $105°$; (iii) 16 lbs. and 21 lbs. at $120°$. *Ans.* (i) 6·56 lbs.   (ii) 3·15 lbs.   (iii) 19·04 lbs.

Ex. 7. If the three forces in Ex. 4 act in directions represented by the sides of an equilateral triangle, taken in order: determine their resultant. *Ans.* A force $3\sqrt{3}$, acting at right angles to the force 6.

Ex. 8. Three forces acting on a particle keep it in equilibrium: the greatest force is 5 lbs., and the least is 3 lbs., and the angle between two of the forces is a right angle: find the other force. *Ans.* 4 lbs.

Ex. 9. Two equal forces act at a certain angle on a particle, and have a certain resultant: also if the direction of one of the forces be reversed and its magnitude be doubled, the resultant is of the same magnitude as before: shew that the two equal forces are inclined at an angle of $60°$.

Ex. 10. Determine the resultant of four forces of 5, 6, 9, 10 lbs. acting on a particle and represented in direction by OA, OB, OC, OD, respectively, where O is the point of intersection of the diagonals of a square $ABCD$.

Ans. $4\sqrt{2}$ lbs., in the direction bisecting the angle COD.

Ex. 11. Forces P, $P\sqrt{3}$, and $2P$ act on a particle: find the angles between their respective directions that there may be equilibrium.

Ans. Between P and $P\sqrt{3}$, $90°$; between P and $2P$, $120°$; between $P\sqrt{3}$ and $2P$, $150°$.

Ex. 12. Five equal forces act on a particle, in directions parallel to five consecutive sides of a regular hexagon taken in order; find the magnitude and direction of their resultant.

Ans. The direction is parallel to the third force, and the magnitude equal that of any one force.

20. The following proposition is sometimes useful.

If two forces acting on a particle be represented by m times the line OA, *and n times the line* OB, *respectively,* their resultant is represented by $(m+n)$ times the line OG, where G is the point between A and B, such that $mAG = nBG$.

By the triangle of forces mOA is equivalent to mGA and mOG, and the force nOB to nGB and nOG. But since $mAG = nBG$, and they are opposite, these two forces counterbalance one another, so that we are left with $(m+n)\,OG$ only.

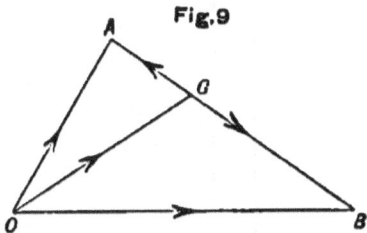

Fig.9

21*. *Def.* Let $A_1, A_2, A_3 \ldots A_n$ be a series of points; join $A_1 A_2$, and take B_1 between them, so that
$$A_1 B_1 = B_1 A_2;$$
join $B_1 A_3$, and take B_2 between them, so that
$$2B_2 B_1 = B_2 A_3;$$
join $B_2 A_4$, and take B_3 between them, so that
$$3B_3 B_2 = B_3 A_4;$$
and so on until we arrive at B_{n-1}: this point is called the Centroid of $A_1, A_2, A_3 \ldots A_n$.

The centroid of the n points $A_1, A_2 \ldots A_n$ is sometimes defined as the point whose distance from any plane is one n^{th} the sum of the distances of $A_1, A_2 \ldots A_n$ from that plane. We can easily shew that the definition of the centroid we have already given leads to this definition also.

Draw $A_1 M_1$, $A_2 M_2$ &c. $B_1 N_1$, $B_2 N_2$ &c. perpendicular to any given plane. Draw $A_1 n_1 m_2$ parallel to $M_1 N_1 M_2$, and $B_1 n_2 m_3$ parallel to $N_1 N_2 M_3$.
$$A_1 M_1 + A_2 M_2 = n_1 N_1 + A_2 m_2 + m_2 M_2$$
$$= 2n_1 N_1 + 2B_1 n_1 = 2B_1 N_1,$$
$$A_1 M_1 + A_2 M_2 + A_3 M_3 = 2B_1 N_1 + m_3 M_3 + A_3 m_3$$
$$= 3n_2 N_2 + 3B_2 n_2 = 3B_2 N_2.$$

This proves the statement for two and three points, and by the method of induction the proof can easily be extended to any number of points.

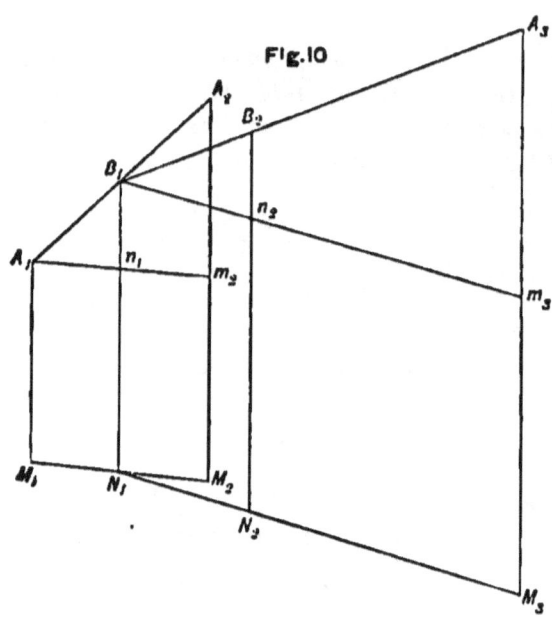

Fig.10

Note. The distances from the plane must be considered positive when they are on one side of it, negative when they are on the other. We may suppose that any number of the points become coincident: for instance, if A_2 and A_3 coincide with A_1, B_1 and B_2 will also coincide with A_1, and B_3 will be in the line A_1A_4, and such that $B_3A_4 = 3B_3A_1$. We may extend the idea of the centroid by supposing that some of the points are negative, in which case the process of finding the centroid will be somewhat modified: for instance, if A_3 be a negative point, B_2 will be in A_3B_1, but beyond B_1 not between B_1 and A_3, and such that $B_2A_3 = 2B_2B_1$: as B_2 is the centroid of two positive and one negative point B_3 will divide the line B_2A_4 equally. Also the distance of a negative point from a plane must be taken of opposite sign to what it would be if the point were a positive one, and in estimating the number of points, we must take the difference between the numbers of positive and negative points.

22*. Prop. *If OA_1, OA_2, OA_3, &c. ...... OA_r represent a number of forces acting on a particle, their resultant will* be represented by r times the line OB_{r-1}, where B_{r-1} is the centroid of $A_1, A_2 \ldots A_r$.

For, by the last proposition, putting $m = n = 1$, the resultant of OA_1 and OA_2 is $2OB_1$; putting $m = 1, n = 2$, that of OA_3 and $2OB_1$ is $3OB_2$, and so on, until we obtain rOB_{r-1} as the final resultant.

After reading Chap. IV. it will be obvious that the centroid of a number of points is the Centre of Mass of equal particles situate one at each point. As a direct result of this proposition we see, that the resultant attraction or repulsion on a particle of any mass of which each particle attracts or repels with a force varying as its distance and its mass conjointly, is the same as the attraction or repulsion of the whole mass collected at its Centre of Mass.

Ex. 1. Find a point such that, if it be acted on by forces represented by the lines joining it to the vertices of a triangle, it will be in equilibrium.

The required point must be the centroid of the three points, i.e. (Art. 21) the point of intersection of the lines drawn from the vertices to the middle points of the opposite sides.

Ex. 2. O is any point in the plane of a triangle ABC, and D, E, F are the middle points of the sides. Shew that the system of forces OA, OB, OC is equivalent to the system OD, OE, OF.

It can be shewn that the centroid of the points A, B, C is also that of D, E, F.

Ex. 3. The circumference of a circle is divided into a given number of equal parts, and forces acting on a particle are represented by straight lines drawn from any point to the points of division: shew that their resultant passes through the centre of the circle, and that its magnitude varies as the distance of the point from the centre.

The centre of the circle is clearly the centroid of the points.

Ex. 4. AOB and COD are chords of an ellipse parallel to conjugate diameters: forces are represented in magnitude and direction by OA, OB, OC, OD: shew that their resultant is represented in direction by the straight line which joins O to the centre of the ellipse, and in magnitude by twice this line.

The centroid of the points A, B, C, D is midway between O and the centre.

Ex. 5. Straight lines are drawn from any point parallel to the four sides of a parallelogram: find the magnitude and direction of the resultant of the forces represented by these four straight lines.

Ans. The direction is along the line joining the point with the centre of the parallelogram, and the magnitude is represented by twice this line.

23. *Def.* The *components* of a force, in two or in three given directions, are the forces which acting in those directions, will have the given force for resultant.

As it is frequently desirable to replace *two or more* forces by *one* (their resultant), so also is it to replace *one* force by *two* (its components), in two given directions in the same plane with it, and sometimes by *three* in three given directions, which are not all in one plane and no two of which are in the same plane as the single force.

For instance, imagine a particle, free to move in a straight groove, to be pulled by a string making an angle with the groove : it is clear that the tendency of the force is twofold, viz. (1) to make the particle move along the groove and (2) to press it against the groove. Also it is clear that the one effect might be produced by a force along the groove and the other by a force at right angles to the groove: these two separate forces will be the components in the corresponding directions of the force exerted by the string.

We have seen that the mechanical problem of compounding two forces into one is the same as the geometrical one of constructing the diagonal of a parallelogram, having given two adjacent sides : so also to resolve one force into its two components in two given directions in its plane, we have to construct the parallelogram, having given one diagonal and lines to which the sides are respectively parallel.

24. *To find the components of a given force in* two *given directions in its plane.*

Let OC represent the force. Draw OA, CB parallel to the line giving one direction, and OB, CA parallel to the line giving the other. By the parallelogram of forces, the force OC is the resultant of OA and OB, which, being in the given directions, are the components required.

We can easily express the magnitudes of these components in terms of OC and the angles the given directions make with OC.

STATICS OF A SINGLE PARTICLE. 23

Let P be the force represented by OC and let the angles COA, COB be α, β.

Then
$$OA : OC = \sin OCA : \sin OAC$$
$$= \sin COB : \sin AOB$$
$$= \sin \beta : \sin (\alpha + \beta);$$

∴ the component in direction $OA = P \dfrac{\sin \beta}{\sin (\alpha + \beta)}$.

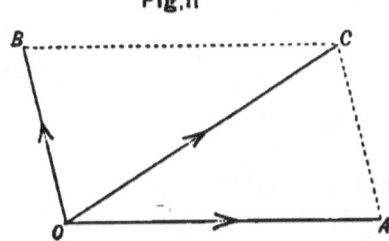

Fig. 11

Similarly that in direction $OB = P . \dfrac{\sin \alpha}{\sin (\alpha + \beta)}$.

Since we can construct any number of parallelograms having a given diagonal, the number of ways in which we can resolve a single force into two is infinite. The most important case is when the two directions along which the resolution takes place are at right angles to one another.

25. *Def.* When the directions of the two components of a force are at right angles to one another, each component is called the *resolved part* of the force in the corresponding direction. When we speak then of the resolved part of a force in any direction, it is understood that the force is resolved into two components, one in the specified direction, and the other in the direction at right angles to it, and in the plane containing this direction and that of the original force.

Let Ox be a given straight line, and let OA_1, OA_2, OA_3...... represent a number of forces P_1, P_2, P_3......, whose directions, which are not necessarily in one plane, make angles θ_1, θ_2, θ_3, &c. with Ox.

Produce xO backwards to x', and draw A_1M_1, A_2M_2, A_3M_3, &c. perpendicular to xOx'. Then OM_1, OM_2, OM_3, &c.

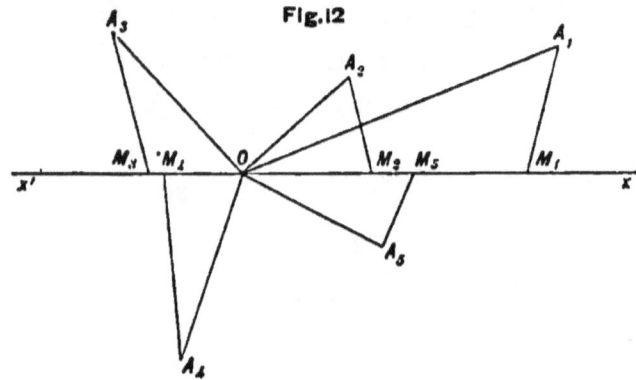
Fig. 12

represent the resolved parts of P_1, P_2, P_3, &c. respectively along Ox, and M_1A_1, M_2A_2, M_3A_3, &c. the resolved parts perpendicular to Ox.

It is found convenient to adopt the convention that forces in direction Ox, from left to right, be considered positive, while those in the opposite direction are considered negative.

In the above figure it will be seen that with this convention the resolved parts along Ox of P_1, P_2 and P_5 are positive, while those of P_3 and P_4 are negative.

$OM_1 = OA_1 \cos \theta_1$, $A_1M_1 = OA_1 \sin \theta_1$, $OM_2 = OA_2 \cos \theta_2$, &c.

Hence the *numerical* values of the resolved parts of the forces along Ox are $P_1 \cos \theta_1$, $P_2 \cos \theta_2$, &c. and those perpendicular to it are $P_1 \sin \theta_1$, $P_2 \sin \theta_2$, &c.

It is easily seen that these values also give the *algebraical* values of the resolved parts, the signs being determined in accordance with the above convention.

Note. When the forces are in one plane, their resolved parts at right angles to Ox are in the same straight line, but not otherwise.

STATICS OF A SINGLE PARTICLE.   25

If X, Y are the resolved parts of a force P, in two directions, making angles θ and $\frac{\pi}{2} - \theta$ respectively, with P, we have seen that

$$X = P \cos \theta, \text{ and } Y = P \sin \theta,$$
$$\therefore P = \sqrt{(X^2 + Y^2)} \text{ and } \tan \theta = Y/X.$$

26*. *To find the components of a force in* three *given directions, which are not all in the same plane, and no two of which are in the same plane as the original force.*

Let the line AB represent the given force.

Through both A and B draw three planes parallel to

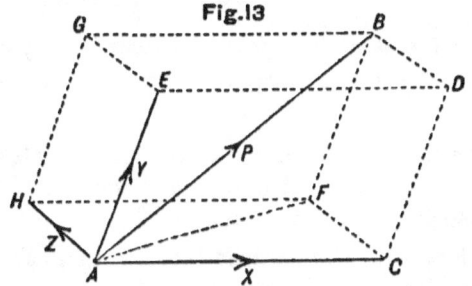

Fig.13

each pair of the given directions. These six planes will form the faces of a parallelopiped of which AB is the diagonal, and each edge of which will be parallel to one of the given directions.

By Art. 13 the edges AE, AC, and AH will represent forces of which AB is the resultant, and which are therefore the required components.

The case which is of most interest is when the given directions are mutually at right angles to one another: the components are then termed the *resolved parts* in the corresponding directions.

Let P be the given force, X, Y, Z the resolved parts in the directions AC, AE, AH respectively, which make angles

a, β, γ with AB. But $AC = AB \cos a$, $AE = AB \cos \beta$, and $AH = AB \cos \gamma$,

and
$$AB^2 = BF^2 + AF^2 = AE^2 + AC^2 + AH^2,$$
$$\therefore X = P \cos a, Y = P \cos \beta, Z = P \cos \gamma,$$

and
$$P^2 = X^2 + Y^2 + Z^2.$$

Hence
$$\cos^2 a + \cos^2 \beta + \cos^2 \gamma = 1.$$

Ex. 1. Show how to resolve a given force into two others, of given magnitude. When is this impossible?

Ex. 2. Find the components of a force P, when they both make angles of $30°$ with it. *Ans.* Each is $\tfrac{1}{3} P \sqrt{3}$.

Ex. 3. Find the components of a force P in two directions, making angles of $60°$ and $45°$ with P on opposite sides.
Ans. $(\sqrt{3}-1) P$ and $\tfrac{1}{2} (3\sqrt{2} - \sqrt{6}) P$.

Ex. 4. Three forces of 5, 2, and 7 lbs. respectively act on a particle in directions mutually at right angles: determine the magnitude of their resultant. *Ans.* $\sqrt{78}$ lbs.

Ex. 5. Three forces, represented by three diagonals of three adjacent faces of a cube which meet, act at a point: shew that their resultant is equal to twice the diagonal of the cube.

Each of the forces may be resolved into two components, represented by those edges of the corresponding face, which meet in the point: the three forces are equivalent then to the three forces represented by twice the edges of the cube, which meet in the point, i.e. to twice the diagonal of the cube. A similar result holds for any parallelopiped.

The *purely geometrical* propositions of the next three Articles are extremely useful.

27. *Def.* If perpendiculars be dropped from the ends of a given finite straight line on any other given straight line, the length intercepted between the feet of these perpendiculars is called the *orthogonal projection* of the first line on the second. (The two lines are not necessarily in one plane.)

Let AB be the given finite line, PQ the line on which it is to be projected.

Draw Aa, Bb perpendicular to PQ, then ab is the orthogonal projection of AB on PQ.

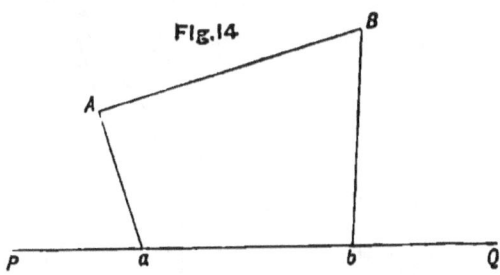

Fig. 14

We shall make a convention here, similar to that we have already made about the resolved parts of forces: viz. if AB be regarded as drawn from A to B, its projection is ab, measured from a to b, whereas if BA be measured from B to A, its projection is ba, measured from b to a. The projections are considered *positive* when measured from *left to right* as ab is, *negative* when measured in the *opposite direction* as ba. These signs apply to figure 14: they are reversed for figure 15.

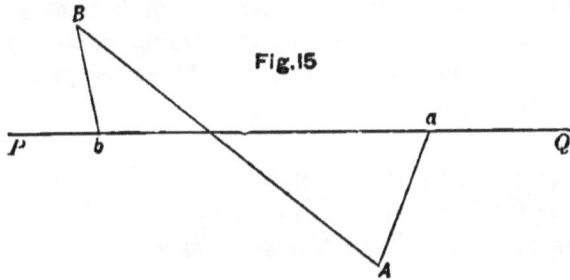

Fig. 15

28. **Prop.** *The orthogonal projection of any line on another is the product of the projected line and the cosine of the angle between them.*

The angle between two lines not in the same plane is the angle between one of them and a line intersecting it and parallel to the other.

Let AB be any finite line, PQ the line on which it is projected, α the angle between them.

Draw Aa perpendicular to PQ, and let $Bb'b$ be a plane through B perpendicular to PQ, cutting the latter in b.

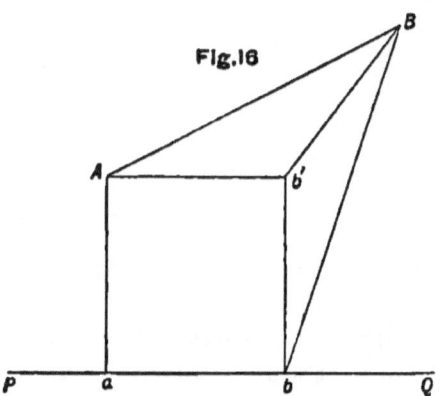

Fig.16

Draw Ab' parallel to ab. Then Ab' is at right angles to the plane $Bb'b$, the angle $Ab'B$ is a right angle, and $BAb' = \alpha$. Since $Aa, b'b$ are both perpendicular to ab, and are in the same plane, they are parallel, and $Ab'ba$ is a parallelogram; hence $ab = Ab' = AB \cos \alpha$.

Observe that α is the angle between AB, and a line drawn from A parallel to PQ in the direction in which the projections are estimated positively. If α is an obtuse angle, the projection is negative. The angle which BA makes with PQ is two right angles greater than that which AB makes with it.

29. Prop. *The algebraical sum of the projections of the two straight lines* AB, BC *on any straight line is equal to the projection of* AC *on the same line.*

Draw Aa, Bb, Cc at right angles to the given straight line PQ, then

$$\text{projection of } AB = ab \text{ (positive)},$$
$$\ldots\ldots\ldots\ldots BC = bc \text{ (negative)},$$
$$\ldots\ldots\ldots\ldots AC = ac \text{ (positive)},$$

and
$$ab - bc = ac,$$

therefore the algebraical sum of the projections of AB, BC = the projection of AC.

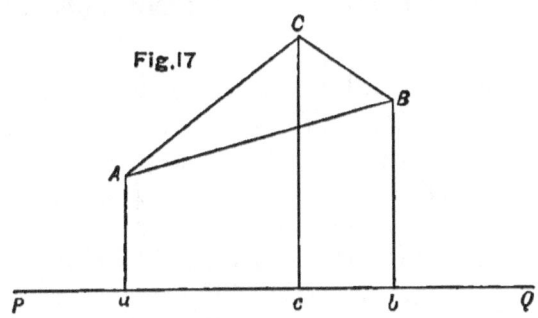
Fig. 17

The above signs refer to the figure given; the student can convince himself of the generality of the truth of this proposition by drawing different figures.

Cor. *The algebraical sum of the projections of the lines* AB, BC, CD, DE, EF, FG (*fig.* 6), *drawn end to end, and measured all the same way round, on any line is equal to the projection of the line* AG.

For the algebraical sum of the projections of AB, BC = the projection of AC,

that of projections of AC, CD = projection of AD,
.... AD, DE = ... AE,
.... AE, EF = ... AF,
.... AF, FG = ... AG,

therefore the algebraical sum of the projections of AB, BC, CD, DE, EF, FG = the projection of AG.

The same holds for any number of such lines.

30. It follows at once from the figure, or from the expressions for the resolved part of a force in any direction, and the projection of a line on a straight line, that the orthogonal projection of the line representing a force, on any straight line, represents in every respect the resolved part of the force in the corresponding direction.

Prop. *The algebraical sum of the resolved parts in any direction of a number of forces acting on a particle, is equal to the resolved part of their resultant in that direction.*

This proposition follows at once from the last, for if (fig. 6) $AB, BC, CD, \ldots FG$ represent the forces, AG represents their resultant; and the algebraical sum of the projections on any straight line, of $AB, BC, \ldots FG$, which projections represent the resolved parts of the forces in the corresponding direction, is equal to the projection of AG, which represents the resolved part of the resultant.

31. We can now obtain expressions for the magnitude and direction of the resultant of a number of given forces acting on a particle.

First, let the directions of the forces all lie *in one plane*.

Let $P_1, P_2 \ldots$ be the forces, whose directions make angles α_1, α_2, &c., with the line Ox in the plane of the forces: let Oy be a line at right angles to Ox in the same plane. Let R be the resultant of the forces and θ the angle its direction makes with Ox.

Then from the proposition just proved

$$P_1 \cos \alpha_1 + P_2 \cos \alpha_2 + \ldots = R \cos \theta,$$

and

$$P_1 \sin \alpha_1 + P_2 \sin \alpha_2 + \ldots = R \sin \theta,$$

therefore,

$$R^2 = [\Sigma(P \cos \alpha)]^2 + [\Sigma(P \sin \alpha)]^2,$$

and

$$\tan \theta = \frac{\Sigma [P \sin \alpha]}{\Sigma [P \cos \alpha]}.$$

32*. Secondly, when the directions of the forces are *not necessarily in one plane*.

Let P_1, P_2, P_3, &c. be the forces, whose directions make with three straight lines Ox, Oy, Oz mutually at right angles, angles $\alpha_1, \beta_1, \gamma_1, \alpha_2, \beta_2, \gamma_2, \alpha_3, \beta_3, \gamma_3$, &c. respectively.

Let R be the resultant of these forces, $\bar{a}, \bar{\beta}, \bar{\gamma}$ the angles its directions make with Ox, Oy, Oz, respectively. Then

$$P_1 \cos \alpha_1 + P_2 \cos \alpha_2 + \ldots = R \cos \bar{a},$$
$$P_1 \cos \beta_1 + P_2 \cos \beta_2 + \ldots = R \cos \bar{\beta},$$
$$P_1 \cos \gamma_1 + P_2 \cos \gamma_2 + \ldots = R \cos \bar{\gamma},$$
$$\therefore R^2 = [\Sigma(P \cos \alpha)]^2 + [\Sigma(P \cos \beta)]^2 + [\Sigma(P \cos \gamma)]^2,$$
$$\cos \bar{a} = \frac{\Sigma(P \cos \alpha)}{\sqrt{\{[\Sigma(P \cos \alpha)]^2 + [\Sigma(P \cos \beta)]^2 + [\Sigma(P \cos \gamma)]^2\}}},$$

with symmetrical expressions for $\cos \bar{\beta}$ and $\cos \bar{\gamma}$.

Conditions of Equilibrium.

33. If the resultant of a number of forces acting on a particle be zero, its resolved part in any direction is zero also; hence

If a system of forces be in equilibrium, the algebraical sum of their resolved parts in any direction is zero.

Conversely, if the algebraical sum of the resolved parts of a number of forces in any direction be zero, the resolved part of their resultant in that direction must be zero also, i.e. the resultant is either zero or acts perpendicularly to that direction. But as a line cannot be perpendicular to each of *two* directions in the same plane as itself, or to each of *three* directions not all in the same plane, the resolved part of a force, which is not zero, cannot be zero in *two* directions in its own plane, or in *three* directions not all in the same plane. Hence

A system of coplanar forces acting on a particle is in equilibrium, provided the algebraical sums of their resolved parts in two directions in the plane are zero. Also if the forces are not in one plane, they are in equilibrium, provided the algebraical sums of their resolved parts in three directions not in the same plane, are severally zero.

These conditions have been directly deduced from the condition that the resultant should be zero: in practice they are often found to be easier of expression than the geometrical one.

Ex. 1. $ABCD$ is a square. A force of 3 lbs. acts along AB, one of 4 lbs. along AC, and one of 5 lbs. along AD; find the magnitude and direction of their resultant.

Ans. $\sqrt{(50+32\sqrt{2})}$, making with AB an angle $\tan^{-1}(7-4\sqrt{2})$.

Ex. 2. Three forces act on a particle in one plane: they are 1 lb., 5 lbs., and 3 lbs. respectively, and the force of 5 lbs. is inclined at an angle of $30°$ to each of the others: find their resultant.

Ans. $\sqrt{(38+20\sqrt{3})}$ lbs., making with the direction of the force of 5 lbs. an angle $\cot^{-1}(5+2\sqrt{3})$ on the side of the force of 3 lbs.

Ex. 3. At the point O the intersection of the diagonals of a square $ABCD$, act forces of 2 lbs. along OA, 4 lbs. along OB, 3 lbs. parallel to CD, and 1 lb. parallel to DA: find their resultant.

Ans. $\sqrt{30}$ lbs., making with CD an angle $\tan^{-1}\tfrac{1}{7}(9+10\sqrt{2})$.

Ex. 4. Three forces P, P and $P\sqrt{2}$ act on a particle in directions mutually at right angles: determine the magnitude of the resultant and the angles between its direction and that of each component.

Ans. $2P$, making with either force P an angle of $60°$, and with $P\sqrt{2}$ an angle of $45°$.

Ex. 5. A particle is placed at the corner of a cube, and is acted on by forces of 1, 2 and 3 lbs. respectively, along the diagonals of the faces of the cube, which meet at the particle: determine the magnitude of the resultant. *Ans.* 5 lbs.

34. *Def.* The *moment* of a force about any *point* is measured by the product of the force into the length of the perpendicular from the point on the line of action of the force.

(The line of action of a force is a line, drawn through the particle on which the force acts, in the direction of the force.)

Let P be a force acting on a particle situate at A, and O be any point: draw OM perpendicular to P's line of action, then $P \times OM$ measures the moment of P about O. The magnitude of the moment of P is clearly independent of the position of A, provided the line of action remain the same.

It is convenient to make the convention that if the

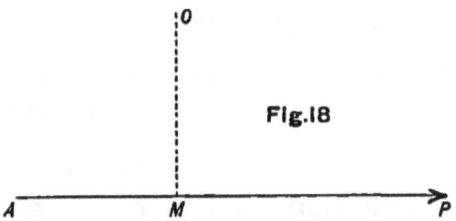

Fig.18

force tends to move the particle round O in the same direction as the hands of a watch, when looked at from above, the moment is of one sign, when in the opposite direction, of the other sign. The latter is generally taken as the positive moment. In the above figure the moment is positive.

The moment of a force is zero, when the force itself is zero, or when its line of action passes through the point about which the moments are estimated, and in these two cases only.

The student is recommended to accept the above definition of the moment of a force, and to follow the theorems concerning it, without troubling himself at first to learn the physical meaning of the term.

35. **Prop.** *The moment of a force about a given point is algebraically equal to the moment of its resolved part at right angles to the line joining the point with the particle, on which the force acts.*

Let P be the force acting on the particle at A, O

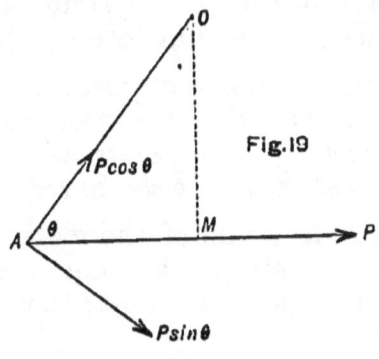

Fig.19

the given point. Draw OM perpendicular to P's line of action and join OA. Let $OAM = \theta$. The resolved part of P at right angles to OA is $P \sin \theta$.

The moment of $P \sin \theta$ about $O = P \sin \theta \cdot OA$

$\qquad\qquad\qquad\qquad\qquad = P \times OM$

$\qquad\qquad\qquad\qquad\qquad =$ moment of P about O.

It is also evident that these moments are of the same sign.

36. **Prop.** *The algebraical sum of the moments of a number of coplanar forces, acting on a particle, about any point in their plane is equal to the moment of their resultant about the same point.*

Let A be the position of the particle, O the given point.

The algebraical sum of the moments of the forces about $O =$ the algebraical sum of the moments about O of their resolved parts perpendicular to OA

$= OA \times$ the algebraical sum of these resolved parts

$= OA \times$ resolved part of their resultant in this direction

$=$ moment of their resultant about O.

Cor. If the forces are in equilibrium, it follows that the algebraical sum of their moments about any point in their plane is zero.

37. By means of the last theorem the sufficient conditions of equilibrium of a system of coplanar forces acting on a particle, can be put into a different form.

Prop. *A system of coplanar forces acting on a particle is in equilibrium, provided the algebraical sum of the moments about each of two points in the plane but not in a straight line with the particle, be zero.*

For the algebraical sum of the moments of the forces about any point in their plane is equal to the moment of their resultant about the same point: therefore the moment of their resultant about each of the two points is zero, so that either the resultant is zero, or its line of action passes through both the points; the latter cannot be the case as the line of action passes through the particle. Hence the forces are in equilibrium.

38. *Def.* If a force be resolved into two components respectively parallel and perpendicular to a given straight line, the product of the latter component into the common perpendicular to its line of action and the given line, is called the *moment* of the force *about the given line*.

If the force tend to turn the particle it acts on, in one direction about the given line, the moment receives the positive sign; if in the opposite direction, the moment is taken to be negative.

Let P be the force acting on a particle at A, CD the

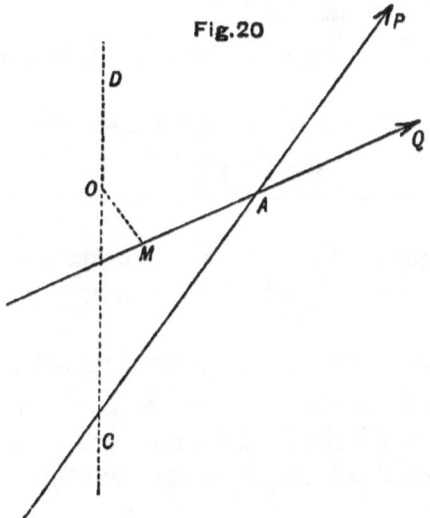

Fig.20

given straight line. Let Q be the resolved part of P at right angles to CD, O the point where CD intersects a plane through A perpendicular to CD: Q's direction is in this plane: draw OM perpendicular to it.

Then P's moment about $CD = Q \times OM$
$\qquad\qquad\qquad\qquad\quad =$ moment of Q about O.

Now OM is perpendicular to CD and Q, and therefore to the plane parallel to CD containing Q's line of action, i.e. to the plane PAQ. But since CD is parallel to this plane, all points in CD are at the same distance (OM) from it; and Q is the same wherever A is in the same line of action; therefore the moment of P about CD is independent of the position of A in its line of action.

It is obvious that the moment of a force about a line is zero, if its line of action and the line are coplanar or if the force is zero, and in these cases only.

39. Prop. *The algebraical sum of the moments of a number of forces acting on a particle, about any straight line is equal to the moment of their resultant about the line.*

Let (fig. 20) A be the position of the particle, CD the given line, and O its intersection with the plane through A perpendicular to it.

The algebraical sum of the moments about CD of the forces
= the algebraical sum of the moments about O of their resolved parts perpendicular to CD
= the moment about O of the resultant of these resolved parts
= the moment about O of the resolved part perpendicular to CD of the resultant of the original forces
= the moment about CD of this resultant.

Cor. Hence if the forces are in equilibrium, the algebraical sum of their moments about any line is zero, for if their resultant is zero, its moment about any line is also zero.

40. *Recapitulation.* We began by shewing from purely geometrical considerations how we can compound the independent velocities of a moving point, into a single resultant velocity, by means of the *Parallelogram of Velocities*. From Newton's *First Law* we obtained a general idea of force as that cause, which, acting on a body, tends

to alter the state of rest or uniform motion in a straight line, which is the condition of all bodies not acted on by force. Newton's *Second Law* defined the direction of a force, and stated that its magnitude is proportional to the change of momentum produced by it in any body after acting on the latter for a certain time; from this and the *Parallelogram of Velocities* we deduced the fundamental Proposition in Statics, the *Parallelogram of Forces*. Then followed other theorems, the *Triangle of Forces*, the *Polygon of Forces*, *Lami's theorem*, &c., modifications of the Parallelogram of Forces, which often enable us to solve Statical Problems more easily than the original proposition. Having shewn that the algebraical sum of the resolved parts in any direction of a number of forces acting on a particle is equal to the resolved part of their resultant in that direction, we obtained expressions for the magnitude and direction of the resultant of a number of forces. From this and because the sole necessary and sufficient condition of equilibrium of a number of forces acting on a particle is that their resultant be zero, we obtained a set of conditions of equilibrium which is often easily applied to the solution of problems. Another important set of conditions of equilibrium we deduced from the proposition that the algebraical sum of the moments of a number of forces, about any straight line, or in the case of coplanar forces, about a point, is equal to the moment of their resultant about the same line, or point.

41. *Tension of a String.* A very common way of transmitting force is by means of a flexible string, rope or chain. Now when a string AB is *stretched* by the application of forces it is a matter of everyday experience that if it be cut at any point P, the two ends on either side of P separate: what then prevented the portion AP from moving before the string was cut? Clearly the force which the other portion PB exerted on it, and similarly the latter was prevented from moving by the force which AP exerted on it. But we shall see in Art. 44 that these forces are equal to one another, and act in opposite directions along the lines joining the two adjacent particles on either side of P, i.e. along the tangent at P to the curve formed by the string; if the string is straight, these forces will act

along it. Either of these forces is called the *tension* at P. If the tensions at all points of the string are the same, we speak in general of the *tension of the string*.

There is a limit to the tension which any given string can exert, and if we try to transmit a force greater than this by means of the string, it will break.

The above remarks apply to rods also if they are stretched, but the tension becomes a *thrust*, if the tendency of the forces on them is to compress them.

42*. *Extensible Strings.* The following experimental law, due to Hooke, gives the relation between the extension of an extensible string or rod, the tension along it, and its natural length, i.e. its length when unstretched. *For strings of the same material and thickness, the extension varies as the tension and the natural length conjointly.*

If l be the natural length, l' the length when stretched, t the tension, we may write the law symbolically,

$$l' - l \propto lt,$$

or
$$l' - l = \frac{lt}{\lambda},$$

where λ is a constant for the particular string in question.

We assume that the tension of the string is t throughout the whole length to which we apply the law. For many substances, such as steel, this law is only true so long as the extension is small compared with the natural length, but in others, such as india-rubber, the limits within which it holds are much wider. It is easily seen that λ is the tension, which, if the law held whatever the extension is, would stretch the string or rod to double its natural length.

λ is termed the *Modulus of Elasticity* for strings of the same material and thickness.

43. When any body is acted on by force we ascribe this force to some other body: e.g. we ascribe the force which causes a body to fall to the ground, to the earth; the forces by which the planets are kept in their orbits we ascribe to the Sun. If we wish to move any body we must act on it by means of some other, such as the hand. Now we know by experience that if we strike a table with the hand the latter is stopped: if we throw a ball against a wall, it rebounds. Since the hand is stopped and the ball rebounds, each must have been acted on by force, i.e. when the hand acted on the table, the table reacted on the hand, and when the ball acted on the wall, the wall reacted on the ball. These

STATICS OF A SINGLE PARTICLE.

illustrations shew that when a body A acts on a body B, A is acted on in return by B. The exact relation between the action and the reaction is expressed by Newton's *Third Law*, which is

To every action there is an equal and contrary reaction.

All forces then occur in pairs, each pair consisting of equal and opposite forces, an *action* and a *reaction*. The mutual action between two bodies is termed the *Stress* between them. It is obvious that all forces are stresses.

We shall often have to consider the equilibrium of bodies which are not free to move in any direction, but are constrained by surfaces, curves, &c. with which they are in contact. For instance, suppose a small body inside a fine tube; the only possible motion of the body is along the tube, i.e. the tube itself will supply the force necessary to prevent motion in any other direction: if then the resultant force on the body, not including the force exerted by the tube, be perpendicular to the tube, we know that the body is in equilibrium. Similarly, if a particle be on a plane, and the resultant force, not including the force exerted by the plane, be perpendicular to the plane and towards it: this force, however great, will be counteracted by the force exerted by the plane and the particle will be in equilibrium. If, however, the resultant force is away from the plane, the particle will move, as the plane cannot exert a force to prevent motion away from itself.

Smooth planes or tubes are those which can only exert forces perpendicular to themselves and are the only ones with which we are concerned at present. A plane or tube which can oppose the motion of a particle along itself, or in other words, can exert a force not entirely perpendicular to itself, is termed *rough*. The same may be said of a curved surface if we take the tangent plane at the point where the particle touches it, as the plane considered above.

Such forces as the pressures exerted by surfaces, &c., and the tensions of inextensible strings, are called into play by the actions of other forces which tend to press the body against the surface, or to stretch the string; the former only act when the latter do. Also, if the surface and the string be supposed strong enough, each is capable of exerting a force of any magnitude, if such a force is necessary to preserve equilibrium. Such forces are termed *Passive* forces, and it is axiomatic that their magnitudes will always adapt themselves so as to maintain equilibrium, if possible.

ILLUSTRATIVE EXAMPLES.

Ex. 1. Assuming that the Parallelogram of Forces holds as regards direction, to prove it as regards magnitude.

Let AB, AC represent two forces in magnitude and direction: complete the parallelogram $ABDC$, and join AD. By hypothesis AD is the direc-

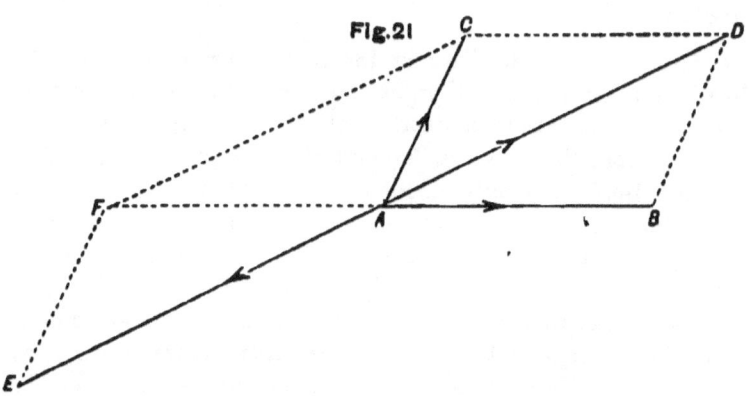

Fig. 21

tion of the resultant of the two forces. The force then which will counteract these two forces must act in the direction DA: produce DA to E so that AE represents the magnitude of this last force. The three forces AB, AC, AE are in equilibrium.

Complete the parallelogram $ACFE$, and join AF. By hypothesis, AF represents the direction of the resultant of AC and AE; AF then is in a straight line with AB; i.e. is parallel to CD, and $ADCF$ is a parallelogram.

$$\therefore AD = FC = AE.$$

Hence AD represents the magnitude of the resultant of the forces represented by AB and AC, since it is equal to AE which represents the force that would counteract them.

The converse proposition could be proved in a similar way.

Ex. 2. Forces P, Q act at a point O, and their resultant is R: if any transversal cut their directions in the points L, M, N respectively, shew that

$$\frac{P}{OL} + \frac{Q}{OM} = \frac{R}{ON}.$$

STATICS OF A SINGLE PARTICLE.   41

Through N draw Nl parallel to OM, and Nm parallel to OL.

The triangle OlN has its sides parallel to the directions of the forces P, Q, R respectively, and if R be reversed, these forces are in equilibrium;

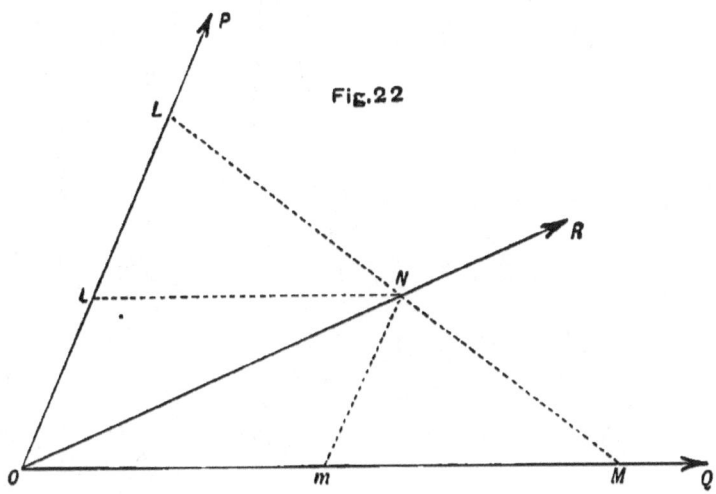

Fig. 22

hence (Art. 16) each side is proportional to the force to whose direction it is parallel, i.e. $Ol = \mu P$, $lN = \mu Q$, $ON = \mu R$.

$$\therefore \frac{P}{OL} + \frac{Q}{OM} = \frac{1}{\mu}\left(\frac{Ol}{OL} + \frac{lN}{OM}\right)$$

$$= \frac{1}{\mu}\left(\frac{Ol}{OL} + \frac{lL}{OL}\right) = \frac{1}{\mu} = \frac{R}{ON}.$$

This result is obtained directly by expressing the fact that the sum of the resolved parts of P and Q perpendicular to LNM is equal to the resolved part of R.

Ex. 3. Shew that the resultant of three forces acting on a particle and represented by AP, PB, PC, where P is the orthocentre of a triangle ABC, is represented in magnitude and direction by the diameter of the circle ABC, which passes through A.

Draw AH the diameter of the circle ABC: join BH, CH: then the angle ABH is a right angle. By the triangle of forces the resultant of AP, PB is represented by AB, since the forces AP, PB, BA acting on a particle would maintain equilibrium.

We have then to prove that AH is the resultant of AB and PC.

But by the triangle of forces, AH is the resultant of AB and BH;

hence the problem reduces to the geometrical one of proving that PC is

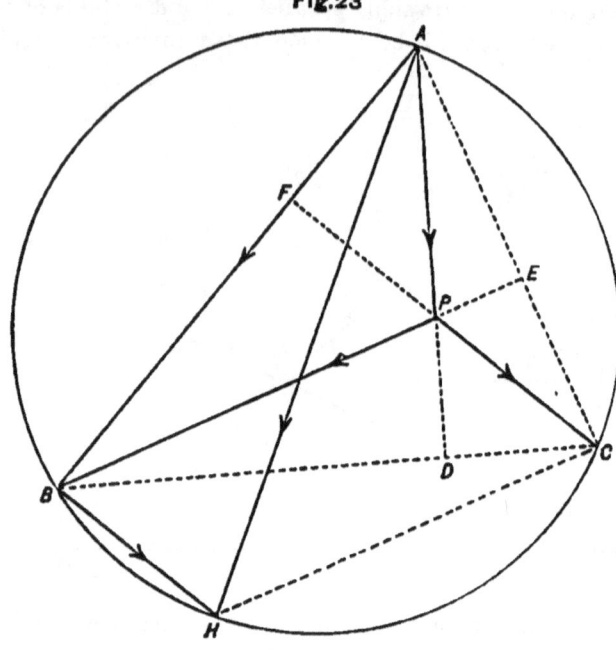

Fig.23

equal and parallel to BH. Since they are both at right angles to AB they are parallel.

For a similar reason, BP and CH are parallel.

∴ $PBHC$ is a parallelogram.

∴ $PC = HB$.

Ex. 4. Two forces act along the sides CA, CB of a triangle ABC, their magnitudes being proportional to $\cos A$, $\cos B$. Prove that their resultant is proportional to $\sin C$, and that its direction divides the angle C into two parts, $\quad \frac{1}{2}(C+B-A)$, $\frac{1}{2}(C+A-B)$.

Let $k \cos A$, $k \cos B$ be the forces, R their resultant, θ the angle its direction makes with CA.

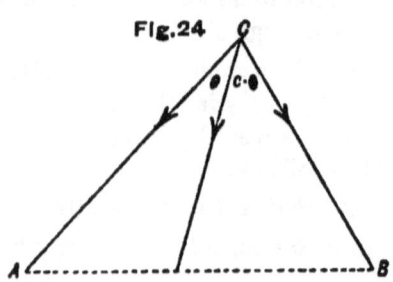

Fig.24

STATICS OF A SINGLE PARTICLE. 43

If R were reversed, the three forces would be in equilibrium (Art. 14), and then each force would be proportional to the sine of the angle between the other (Art. 18).

$$\therefore R : k \cos A : k \cos B = \sin C : \sin (C - \theta) : \sin \theta.$$

$$\therefore \frac{\sin (C - \theta)}{\sin \theta} = \frac{\cos A}{\cos B},$$

solving for θ we obtain $\cot \theta = \tan B$,

$$\therefore \theta = \tfrac{1}{2}\pi - B = \tfrac{1}{2}(A + C - B),$$

and
$$C - \theta = \tfrac{1}{2}(C + B - A),$$

and
$$R = \frac{k \cos B \sin C}{\sin \theta} = k \sin C.$$

Ex. 5. Three forces P, Q, R in one plane, act on a particle, the angles between R and Q, P and R, and P and Q being α, β, and γ respectively: prove that their resultant

$$= \{P^2 + Q^2 + R^2 + 2QR \cos \alpha + 2RP \cos \beta + 2PQ \cos \gamma\}^{\frac{1}{2}}.$$

Let X_1, Y_1 be the resolved parts of P in two directions at right angles to one another, X_2, Y_2 and X_3, Y_3 those of Q and R respectively in the same directions. Then (Art. 31) the resultant

$$= \sqrt{\{(X_1 + X_2 + X_3)^2 + (Y_1 + Y_2 + Y_3)^2\}}.$$

But $(X_1 + X_2)^2 + (Y_1 + Y_2)^2 = $ (resultant of P, $Q)^2$
$$= P^2 + Q^2 + 2PQ \cos \gamma.$$

Similarly $(X_2 + X_3)^2 + (Y_2 + Y_3)^2 = Q^2 + R^2 + 2RQ \cos \alpha$

and $(X_3 + X_1)^2 + (Y_3 + Y_1)^2 = P^2 + R^2 + 2PQ \cos \beta.$

$\therefore$ adding

$(X_1 + X_2 + X_3)^2 + (Y_1 + Y_2 + Y_3)^2 + X_1^2 + Y_1^2 + X_2^2 + Y_2^2 + X_3^2 + Y_3^2$
$$= 2(P^2 + Q^2 + R^2 + PQ \cos \gamma + PR \cos \beta + QR \cos \alpha),$$

$\therefore (X_1 + X_2 + X_3)^2 + (Y_1 + Y_2 + Y_3)^2$
$$= P^2 + Q^2 + R^2 + 2PQ \cos \gamma + 2QR \cos \alpha + 2PR \cos \beta,$$

$\because P^2 = X_1^2 + Y_1^2$, $Q^2 = X_2^2 + Y_2^2$, $R^2 = X_3^2 + Y_3^2$:

whence the required result.

The same result can be obtained by resolving in three directions mutually at right angles, when P, Q, R are not in one plane.

Ex. 6. Forces act through the angular points of a triangle perpendicular to the opposite sides, and are measured by the cosines of the corresponding angles; shew that their resultant is $\sqrt{(1 - 8 \cos A \cos B \cos C)}$.

We obtain the resultant by substituting in the last example
$\cos A$ for P, $\cos B$ for Q, $\cos C$ for R,
$\pi - A$ for α, $\pi - B$ for β, and $\pi - C$ for γ.

∴ the square of the resultant $= \cos^2 A + \cos^2 B + \cos^2 C$
$\qquad - 2\cos A \cos B \cos C - 2\cos B \cos C \cos A - 2\cos C \cos B \cos A$
$\quad = 1 - \sin^2 A + \cos^2 B + \cos^2 C - 6\cos A \cos B \cos C$
$\quad = 1 + \cos(A-B)\cos(A+B) + \cos^2 C - 6\cos A \cos B \cos C$
$\quad = 1 - \cos C\{\cos(A-B) + \cos(A+B)\} - 6\cos A \cos B \cos C$
$\quad = 1 - 8\cos A \cos B \cos C.$

Ex. 7. Prove that the resultant of forces 7, 1, 1, and 3 acting from one angle of a regular pentagon towards the other angles, taken in order, is $\sqrt{71}$.

Let $ABCDE$ be the pentagon, AB, AC, AD, AE the lines along which the forces 7, 1, 1, and 3 respectively act. Draw AF at right angles to DC. The angles BAF, EAF each $= 54°$, the angles CAF, DAF each $= 18°$.

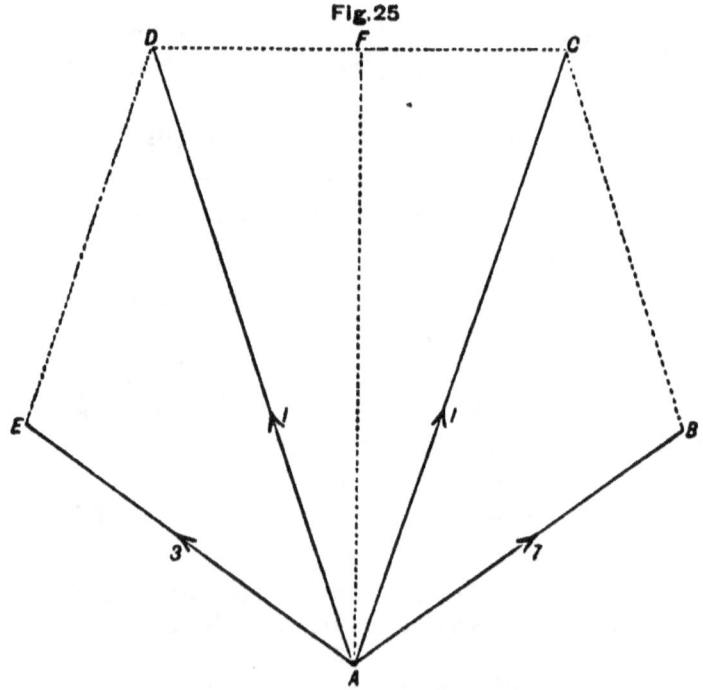

Fig. 25

Resolve the forces in directions AF, FC.

STATICS OF A SINGLE PARTICLE.   45

X, the algebraical sum of resolved parts in direction AF'
$$= (7+3)\cos 54^{0} + (1+1)\cos 18^{0} = 10\cos 54^{0} + 2\cos 18^{0}$$
$$= \tfrac{1}{2}\{5\sqrt{(10-2\sqrt{5})} + \sqrt{(10+2\sqrt{5})}\}.$$

Y, the algebraical sum of resolved parts in direction FC
$$= (7-3)\sin 54^{0} + (1-1)\sin 18^{0} = 4\sin 54^{0} = \sqrt{5}+1.$$

Whence the resultant, $\sqrt{(X^2+Y^2)} = \sqrt{71}$.

The student has sometimes a difficulty in choosing the lines along which he should resolve the forces, since all directions are open to him for that purpose: it is very important that he should select them judiciously, in order that the work may be simplified. The directions selected in the above example were chosen because they were symmetrically placed as regards the forces.

Ex. 8. Prove that if O be the centre of the circumscribing circle, and O' the centre of perpendiculars of a triangle ABC, the resultant of forces represented by OA, OB, OC is represented by OO'.

By Art. 22, we shall prove the required result, by proving that the centroid of A, B, C is in OO', at a distance from O, $\tfrac{1}{3}$ that of O' from O.

Draw OD, $AO'D'$ perpendicular to BC: join AD, cutting OO' in P.

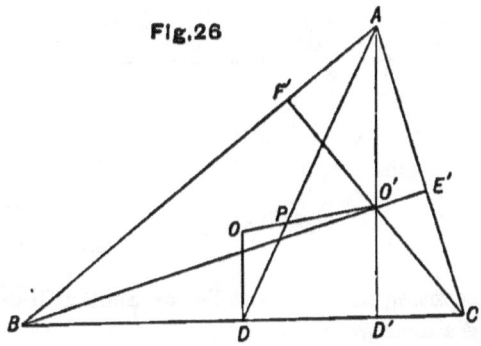

Fig. 26

Now
$OD = R\cos A$, and $AO' = AE'\sec D'AC = AB\operatorname{cosec} C\cos A = 2R\cos A$,
where R is the radius of the circle ABC.

$$\therefore AO' = 2OD \text{ and } O'P = 2OP, \text{ and } AP = 2PD.$$

Hence P is the centroid of ABC, and $OP = \tfrac{1}{3}OO'$.

Ex. 9. A given number of forces acting on a particle are represented in magnitude and direction by straight lines drawn from the focus of a conic to the curve: shew that if the sum of the forces be constant, the locus of the extremity of the line representing the resultant is a straight line.

Let S be the focus; let $SP, SQ \ldots$ be n straight lines drawn from S to the conic so that $SP + SQ + \ldots = $ a constant.

Draw PM, QN, &c. perpendicular to the directrix corresponding to S; then since $SP = ePN$, $SQ = eQN$, &c.,
$$PM + QN + \&c. = \text{a constant.}$$

Let O be the centroid of P, Q, &c., then (Art. 21) the distance of O from the directrix $= \dfrac{PM + QN + \ldots}{n} = $ a constant.

Hence O lies on a straight line parallel to the directrix, and the end

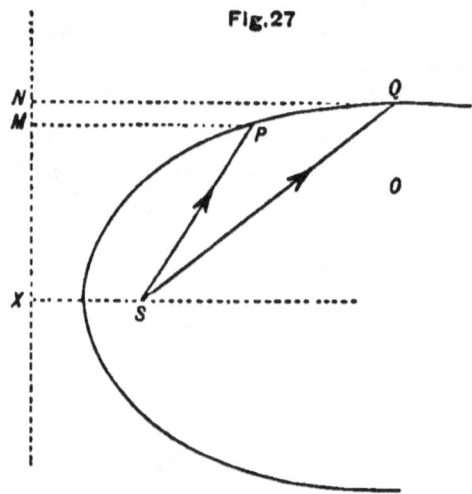

Fig. 27

of the line representing the resultant lies on another line parallel to this, but n times its distance from the focus.

Ex. 10. Forces P, Q, R act from the angular points of a triangle ABC, perpendicular to the opposite sides: prove that if their resultant pass through the centre of the circumscribing circle,
$$P(c \cos B - b \cos C) + Q(a \cos C - c \cos A) + R(b \cos A - a \cos B) = 0.$$

Let O be the orthocentre, O' the centre of the circumscribing circle.

Let D, E, F be the feet of the perpendiculars from O on the sides of the triangle; D', E', F' the feet of the perpendiculars from O' on the same.

Since the resultant of P, Q, R passes through O', its moment about O' is zero.

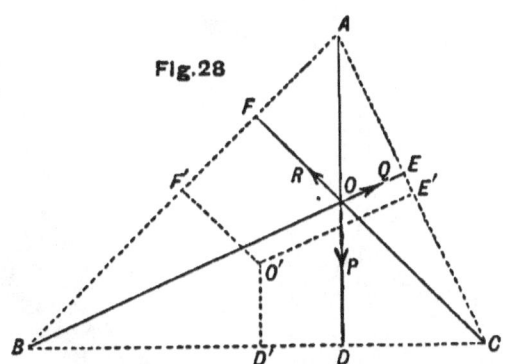

Fig. 28

∴ the algebraical sum of the moments of P, Q, R about O' is zero (Art. 37).

$$\therefore P \cdot DD' + Q \cdot EE' - R \cdot FF' = 0,$$
$$\therefore P(c\cos B - \tfrac{1}{2}a) + Q(a\cos C - \tfrac{1}{2}b) - R(a\cos B - \tfrac{1}{2}c) = 0;$$
$$P(c\cos B - b\cos C) + Q(a\cos C - c\cos A) + R(b\cos A - a\cos B) = 0.$$

In the above figure we see that the forces P and Q tend to move a particle situate at O in the opposite way round O' to that in which R would move it: their moments therefore are of the opposite sign to that of R.

The student may verify for himself that the same result would be obtained were the figure different. He should specially notice in the above example that the required result was obtained by expressing that the algebraical sum of the moments about O' was zero, O' being on the line of action of the resultant.

Ex. 11. A particle of weight W is supported on a smooth inclined plane by means of two strings of given lengths, which are attached to the particle C and to fixed points A, B in a horizontal line in the plane and at a given distance apart. It is required to find the tensions of the strings.

The sides of the triangle ABC being known, the angles which AC, BC make with the horizontal line AB are known: let θ, θ' be their complements. Let a be the inclination of the plane to the horizon. Draw CD perpendicular to AB.

The particle is in equilibrium under the action of four forces, its weight W which acts vertically downwards, the tension T of the string AC, T', that of BC, and the pressure of the inclined plane R, which acts at right angles to the plane.

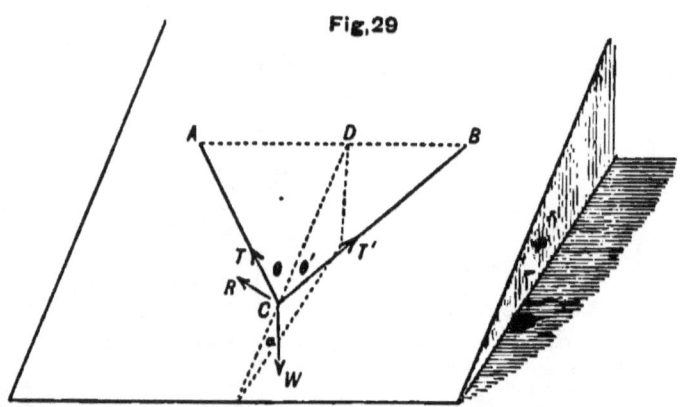

Fig. 29

We shall apply the conditions of equilibrium obtained in Art. 33.

Since the algebraical sum of the resolved parts of the forces in any direction is zero, those in the directions AB and CD must be zero.

$$\therefore T' \sin \theta' - T \sin \theta = 0 \quad \text{(i)},$$
$$T \cos \theta + T' \cos \theta' - W \sin \alpha = 0 \quad \text{(ii)}.$$

R occurs in neither equation, because its direction is perpendicular to all lines in the inclined plane, and W does not occur in the first, because its direction is perpendicular to AB. The inclination of CD to the vertical is the same as that of the plane, and is therefore $\frac{\pi}{2} - \alpha$, so that the resolved part of W along CD is $-W \sin \alpha$.

From (i) and (ii) we obtain

$$T = \frac{W \sin \theta'}{\sin (\theta + \theta')}, \qquad T' = \frac{W \sin \theta}{\sin (\theta + \theta')}.$$

R can be obtained by equating to zero the sum of the resolved parts in the direction perpendicular to the plane, we have then

$$R - W \cos \alpha = 0,$$
or
$$R = W \cos \alpha.$$

The advantages derived from resolving in the particular directions chosen above, are obvious.

Examples on Chapter I.

1. ABC is a triangle: D, E, F are the middle points of the sides BC, CA, AB respectively: shew that forces acting on a particle and represented by the straight lines AD, BE, CF will maintain equilibrium.

2. A, B, C are three points on the circumference of a circle: forces act along AB and BC inversely proportional to these straight lines in magnitude; shew that their resultant acts along the tangent at B.

3. Two forces P and Q have a resultant R which makes an angle a with P: if P be increased by R while Q remains unchanged, shew that the new resultant makes an angle $\frac{1}{2}a$ with P.

4. The resultant of two forces P, Q, acting at an angle θ is equal to $(2m+1)\sqrt{(P^2+Q^2)}$: when they act at an angle $\frac{1}{2}\pi - \theta$, it is equal to $(2m-1)\sqrt{(P^2+Q^2)}$: shew that $\tan\theta = (m-1)/(m+1)$.

5. Compare in terms of the sides of a triangle ABC the forces which acting from O, the centre of the inscribed circle, along OA, OB, OC will balance.

6. Two forces P and $P\sqrt{2}$ act on a particle lying on a smooth horizontal plane. If P makes an angle of $45°$ with the horizon, find the direction of $P\sqrt{2}$ in order that the particle may be in equilibrium.

7. Find within a quadrilateral a point, such that if it be acted on by forces represented by the lines joining it to the angular points of the quadrilateral, it will be in equilibrium.

8. ABC is a triangle, P any point in BC. If PQ represent the resultant of the forces represented by AP, PB, PC, the locus of Q is a straight line parallel to BC.

9. A heavy particle is attached to one end of a string, the other end of which is fixed. Find the force making an angle of $30°$ with the horizontal which must be applied to the particle in order that the string may deviate by an angle of $45°$ from the vertical, and find also the tension of the string.

10. Two forces P, Q act at a point O along two straight lines making an angle a with each other, and have a resultant R: two other forces P', Q' acting along the same two lines have a resultant R'; shew that the directions of R and R' will also include an angle a if

$$PP' + QQ' + 2PQ'\cos a = 0, \text{ or } PP' + QQ' + 2P'Q\cos a = 0.$$

G.

11. If from O, the centre of the circle inscribed in the triangle ABC, forces $\lambda \cos \frac{1}{2}A$, $\lambda \cos \frac{1}{2}B$ act along OB, OA, prove that the magnitude of the necessary force towards C, in order that the resultant may pass through the middle point of AB, is $\lambda \cot \frac{1}{2}C$.

12. A small ring slides on a smooth arc of a circle and rests in equilibrium under the repulsion of three forces P, Q, R, directed from points dividing the circumference into three equal parts: if its position of equilibrium lie on the smaller arc between the points from which the forces Q, R are directed, shew that the pressure exerted by the circle is

$$\{P^2 + Q^2 + R^2 - QR + RP + PQ\}^{\frac{1}{2}}.$$

13. Two particles of weights P and Q respectively, are connected by a string which lies on a smooth circle fixed in a vertical plane; shew that if $\frac{1}{2}\pi$ be the angle subtended at the centre by the string, the inclination of the chord joining P, Q to the horizontal in the position of equilibrium is

$$\tan^{-1} (P \sim Q)/(P + Q).$$

14. OA, OB, OC... are any number of fixed straight lines drawn from a point O, and spheres are described on OA, OB, OC... as diameters. Any straight line OX is drawn through O and a point P taken on it so that OP is equal to the algebraical sum of the lengths intercepted on OX by the spheres. Find the locus of P.

15. Two constant equal forces act at the centre of an ellipse parallel to the directions SP and PH, where S and H are the foci and P is any point on the curve. Shew that the extremity of the line which represents their resultant lies on a circle.

16. Forces are represented by the perpendiculars from the angles of a triangle ABC on the opposite sides: shew that if their resultant passes through the centre of the nine-point circle, the triangle must be isosceles.

17. Three equal forces act at the orthocentre of a triangle ABC, each perpendicular to the opposite side: prove that if the magnitude of each force be represented by the radius of the circle ABC, the magnitude of the resultant will be represented by the distance between the centres of the inscribed and circumscribed circles.

18. The resultant R of any number of forces P_1, P_2, P_3, &c. is determined in magnitude by the equation

$$R^2 = \Sigma (P^2) + 2\Sigma P_r P_s \cos (P_r P_s),$$

where (P_r, P_s) denotes the angle between the directions of P_r, P_s.

19. $ABCDEF$ is a regular hexagon, and at A forces act represented in magnitude and direction by $AB, 2AC, 3AD, 4AE, 5AF$; shew that the length of the line representing their resultant is $\sqrt{351}.AB$.

20. Two small smooth rings of weights W and W', connected by a string, slide upon two fixed wires, the former of which is vertical, and the other inclined at an angle a to the horizon. A weight P is tied to the string, prove that in the position of equilibrium

$$\cot \theta : \cot \phi : \cot a = W : P + W : P + W' + W,$$

where θ, ϕ are the angles which the two portions of the string make with the vertical.

21. $ABCD, A'B'C'D'$ are two parallelograms; prove that forces acting at a point proportional to and in the same direction as $AA', B'B, CC', D'D$, will be in equilibrium.

22. A particle is acted upon by a number of centres of force, some of which attract and some repel, the force being in all cases proportional to the distance, and the intensities for different centres being different: shew that the resultant force passes through a fixed point for all positions of the particle, and examine the one apparent exception.

23. From any point within a regular polygon perpendiculars are drawn on all the sides: shew that the direction of the resultant of all the forces represented by these perpendiculars passes through the centre of the polygon, and find its magnitude.

24. Two heavy rings slide on a wire in the shape of an ellipse whose major axis is vertical, and are connected by a string which passes over a smooth peg at the upper focus: shew that if the weights are equal and the length of the string is equal to that of the axis major, there are an infinite number of positions of equilibrium.

25. Four particles A, B, C, D are attached to the ends of strings whose other ends are tied in a knot at O. Any two particles repel one another with a force which varies directly as the distance and the product of their masses. Shew that when the system is in equilibrium, the volumes of the tetrahedra $OBCD, OCDA, ODAB, OABC$ are proportional to the masses of A, B, C, D respectively.

26. In an ellipse a polygon $PQRS$, &c. is described so that the triangles formed with a side as base and the centre of the ellipse as vertex are of equal area. If O be any point in the plane of the ellipse, prove that the line of action of the resultant of the forces represented by OP, OQ, OR, &c. passes through the centre of the ellipse.

27. Two small heavy rings slide on a smooth wire, in the shape of a parabola, whose axis is horizontal: they are connected by a light string which passes over a smooth peg at the focus: shew that in the position of equilibrium, their depths below the axis are proportional to their weights.

28. Forces P, Q, R act in the lines DA, DB, DC and their resultant meets the plane ABC in G, shew that

$$P/AD : Q/BD : R/CD :: \triangle BGC : \triangle CGA : \triangle AGB.$$

If their resultant be parallel to the plane ABC, then

$$P.DB.DC + Q.DC.DA + R.DA.DB = 0.$$

29. O is any point on the circle circumscribing a triangle ABC, and OL, OM, ON are the perpendiculars from O on the sides. The line LMN meets the perpendiculars from A, B, C on the opposite sides in P, Q, R respectively. Prove that if forces act at O represented by OL, OM, ON, OP, OQ, OR their resultant is represented by $3OK$, where K is the orthocentre.

30. ABC is a triangle and $O_1O_2O_3$ are the centres of the three escribed circles opposite to A, B, C respectively. At any point P, forces act along PO_1, PO_2, PO_3 represented in magnitude by $PO_1.BC$, $PO_2.CA$, $PO_3.AB$, respectively. Shew that if their resultant is of constant magnitude, the locus of P is a circle concentric with the circle circumscribing the triangle $O_1O_2O_3$.

31. A weight attached by a cord to a fixed point O, rests against a frictionless curve in the same vertical plane with O: shew that (1) if the pressure on the curve is to be independent of the position of the weight on it, the curve must be a circle: (2) if the tension in the cord is to be independent of the position of the weight, the curve must be a conic section with O as focus.

32. Two equal particles are connected by a fine string, the particles and string being in a fine smooth elliptic tube, whose semi-circumference is equal to the length of the string. The particles are acted on by constant repulsive forces from one focus: prove that, if these forces are equal, the particles will be in equilibrium in any position in which the string is tight, and if they are unequal, in only one such position.

CHAPTER II.

STATICS OF SYSTEMS OF PARTICLES.

44. WHEN a body composed of a number of particles is in equilibrium, each of these particles is in equilibrium also, and the forces which act on it must therefore satisfy the conditions of equilibrium. But among the forces acting on a particle must be included, not only what are called *External* forces, such as the force of gravity, the pressure and tensions due to other bodies, but also *Internal* forces, i.e. the forces of attraction and repulsion that exist among the different particles composing the body. These forces are by no means always the same in the same body: for example, it is plain that if we try to stretch a rod, the forces that the different particles composing the rod, exert one on another, are different from what they are when we try to compress it. In the former case, the external forces tend to separate particles arranged along a line parallel to the rod's length, in the latter they tend to move them nearer together. To resist these quite opposite tendencies, different internal forces must be called into play. Concerning these internal forces we know by Newton's Third Law that if the particle A exerts on the particle B a force R, (the *action*) in a certain direction, it is itself acted on by a force R, (the *reaction*) in the exactly opposite direction, and also in the same straight line, so that the line of action of each of these forces must be the line joining A and B.

Necessary Conditions of Equilibrium for any body.

45. Without any further assumption about the internal forces that are exerted when any body is in equilibrium, we can determine conditions which must be satisfied by the external forces in such a case.

Since the algebraical sum of the resolved parts in any direction of the forces, which act on each particle of a body in equilibrium, is zero, that of the resolved parts in any direction of all the forces, external and internal, acting on all the particles, is zero also. But as the resolved part of any action is numerically equal, but of opposite sign, to that of the corresponding reaction, the algebraical sum of the resolved parts in any direction of all the internal forces vanishes separately, for the internal forces consist entirely of pairs of forces, equal and opposite to one another. Hence the algebraical sum of the resolved parts of the remaining forces, the external ones, is zero.

Cor. A system of forces keeping a number of particles in equilibrium will, if applied to a single particle, keep it in equilibrium, since the conditions of Art. 33 are satisfied.

46. In a similar way we can shew that the algebraical sum of the *moments about any line*, of the external forces acting on a body in equilibrium, is zero. We have only to substitute '*moments about any line*' for '*resolved parts in any direction*' and the above proof holds.

We may state then that

If any body be in equilibrium under the action of external and internal forces, the algebraical sums both of the resolved parts in any direction, and of the moments about any line, of the external forces, are zero.

If the lines of action of the external forces be *in one plane*, the algebraical sum of their moments *about any*

STATICS OF SYSTEMS OF PARTICLES. 55

point in that plane is zero, being equal to the algebraical sum of their moments about a line through the point in question, and perpendicular to the plane.

47. It is to be noticed that what we have called internal forces are only so *relatively*—the force which is exerted on the particle A by the particle B is an *internal* one, when we are considering a body or system of bodies containing both particles, whereas if B is not contained in the system, the force is an *external* one. It is then very necessary, in applying the above conditions of equilibrium to a system of particles, to know which forces are external and which internal. The force which is an internal one when we are considering the whole body may become an external one, when only a portion of the body is under consideration.

Ex. 1. A picture weighing 10 lbs. is supported by a string which passes over a smooth peg, and has its two ends fastened to the picture: if the tension of the string be 10 lbs., shew that each string makes an angle of 60° with the vertical.

Apply Art. 45, choosing the vertical and horizontal as the directions along which to resolve.

Ex. 2. A rod is supported by means of two strings which are attached to a fixed point, and one to each end of the rod. Assuming that the weight of the rod acts at its middle point, prove that the tensions of the strings are proportional to their lengths.

Apply Art. 46, taking moments about the middle point of the rod.

Ex. 3. A rod of weight W, is supported at an angle of 60° with the horizon by means of strings attached to its ends, the one attached to the upper end making an angle of 60° with the horizon, but in an opposite direction to the rod: find the tensions of the two strings and the inclination of the second to the horizon, assuming that the weight of the rod acts at its middle point.

Ans. $\frac{1}{4}W\sqrt{3}$, $\frac{1}{4}\sqrt{21}W$, the latter acting at an angle $\tan^{-1} 3\sqrt{3}$, to the horizon.

Apply Arts. 45, 46, choosing an end of the rod as the point about which to take moments, and the horizontal and vertical as the directions in which to resolve.

56   STATICS.

Ex. 4. A square $ABCD$, is in equilibrium under the action of four forces, one of 3 lbs. acting along AB, one of 2 lbs. along BC, and one of 3 lbs. along CD; find the magnitude and line of action of the remaining force.

Ans. A force of 2 lbs., acting in direction CB, at a distance equal to $\frac{3}{4}AB$ from BC.

Apply Arts. 45, 46, resolving along AB, BC respectively, and taking moments about B.

Ex. 5. AB is a straight weightless rod, 15 feet long; 4 lbs. is hung at A, 1 lb. at a point 3 feet from A, and a force of 11 lbs. acts vertically upwards at a point 8 feet from B; find what weight must be attached to the rod to maintain equilibrium, and where it must be placed.

Ans. 6 lbs., 2 ft. 8 inches from B.

Ex. 6. Three forces acting at the corners of a triangle, each perpendicular to the opposite side, keep the triangle in equilibrium: prove that each force is proportional to the side to which it is perpendicular.

Take moments about two of the angular points.

Ex. 7. If three forces P, Q, R, acting along the bisectors of the angles of a triangle, at the angular points A, B, C, respectively, keep the triangle in equilibrium: prove that

$$P : Q : R = \cos \tfrac{1}{2}A : \cos \tfrac{1}{2}B : \cos \tfrac{1}{2}C.$$

Take moments about two of the angular points.

48. We have (Art. 45) found *necessary* conditions of equilibrium for any body or system of bodies whatsoever, including liquids, flexible strings, &c.

We shall hereafter (Arts. 52, 54 and 55) find *sufficient* conditions of equilibrium for *rigid* bodies.

Def. When the particles which compose a body, always form the same configuration, or in other words, when the body always retains the same shape and size, whatever forces be applied to it, the body is said to be *rigid*.

We have no experience of bodies, which answer this description perfectly, but we know of many substances which answer it more or less approximately: i.e. we know of many substances, which will submit to the action of considerable forces without undergoing any appreciable change in shape or size. The results which we shall prove *absolutely* true for perfectly rigid bodies, will be so *approximately* for bodies that are approximately rigid.

49. We have already seen that any system of forces acting on a particle is equivalent to a single force, i.e. there is a single force, such that its effect on the particle could not be distinguished from that of the combined forces. The question now presents itself, whether this is so or not when the forces do not all act on a single particle, but on different particles of a system. If the particles form a rigid body, we shall see that under certain circumstances there exists a force, which together with the given forces would keep the body in equilibrium, so that the effect of this force reversed on the body as a whole, is the same as that of the original forces. But it must be remembered that it is only on the body *as a whole* that the effects are the same necessarily: the internal forces called into play by the single force, are not necessarily the same as those called into play by the system of forces, in fact are generally very different. When we cannot find a single force whose effect on the body as a whole is the same as that of the system of forces, we can always find a different system of forces whose effect will be the same, though they will not generally give rise to the same internal forces. It is usual to speak of the single force, when such a one exists, as the *resultant* of the original forces, and the second set of forces as *equivalent* to the first, though it must always be understood that they are so, strictly speaking, *only in one sense*. Even when the body is not rigid, a single force, or set of forces, which would, if the body were rigid, be equivalent in the above sense is said to be, one the resultant of, the other equivalent to, the original system of forces.

50. Prop. *Any rigid body, under the action of any system of forces, can be fixed by applying single forces of the*

requisite magnitude at each of any three *given points of the body not in the same straight line, the direction of the force at one point being at right angles to the plane containing the three points, and that of the force at a second point at right angles to the line joining it to the third.*

Let A, B and C be any three points of the body. The body can be fixed by the following constraints: imagine a very small spherical socket to be made in the body at A, and a ball just smaller than the socket to be placed in it, and the ball to be fixed. Now imagine a very small hoop with its plane perpendicular to AB, to be fixed round B, and also some obstacle to be placed to prevent C from moving at right angles to the plane ABC. The first constraint prevents the body from moving in any way except by turning about A, and exerts a single force through A as the ball and socket touch in only one point: the second prevents B from turning about A, and therefore from moving at all, so that the body can now only turn about AB; the second force acts at right angles to AB. The third constraint prevents C from moving round AB, and therefore from moving at all, and exerts a single force through C at right angles to the plane ABC. As C cannot turn about AB, it is clear that the body is now fixed.

51. This proposition can be extended to the case in which any of the points A, B, and C are not situate in the body. For we may imagine them to be made so in effect, without introducing any forces external to the whole system, by arranging a system of rigid rods, without weight, rigidly connecting them with the body.

Cor. *If the lines of action of the external forces all lie in one plane, the body can be fixed by the application of single forces at any* two *given points A, B in that plane, each force being in the plane, and the direction of one being at right angles to the line AB.*

For by the last proposition, if a third point C be taken in the plane but not in AB, the body can be fixed by the application of suitable forces P, Q, R at A, B, C respectively, the direction of R being perpendicular to the plane, and that of Q perpendicular to AB. The body is now in equilibrium under the action of the internal forces, the original external forces, and the forces of constraint P, Q, R. Hence (Art. 46) the algebraical sum of the moments about AB of all the external forces, including P, Q, and R, is zero: but each of these moments except that of R is zero, since each of the corresponding forces either intersects AB or is parallel to it. The moment of R must therefore be zero, i.e. R itself is zero, since R neither meets AB nor is parallel to it. Similarly we may shew that the moment of Q about every line through A in the plane is zero, i.e. Q is either zero, or lies in the plane in question. Also by taking moments about lines through B in the plane, except AB, we may shew that either P is zero, or it lies in the plane.

52. Prop. *A number of forces acting on a rigid body, their lines of action all being in the same plane, will keep it in equilibrium, provided any of the following sets of conditions hold:*

(1) *If the algebraical sum of their moments about each of three given points in the plane, but not in the same straight line, be zero.*

(2) *If the algebraical sum of their moments about one given point in the plane, and of their resolved parts in any two given directions in the plane, be zero.*

(3) *If the algebraical sum of their moments about two given points in the plane, and of their resolved parts in any given direction in the plane, not at right angles to the line joining the two points, be zero.*

(1) Let A, B, C be the three given points: if the body is not in equilibrium, it can be fixed by applying forces of constraint P and Q at A and B respectively, both in the plane of the forces and Q perpendicular to AB (Art. 51). The whole system of forces *including P and Q* must satisfy the necessary conditions of equilibrium:

therefore the algebraical sum of their moments about A is zero: but the algebraical sum of the moments about A of the forces excluding P and Q is zero, and the moment of P about A is zero also; hence the moment of the remaining force Q is zero, i.e. Q itself is zero, as it does not pass through A. Similarly we can shew that the moments of P about both B and C are zero; hence either P is zero, or it passes through both B and C. As A, B, and C are not in a straight line, P must be zero. Hence the body is in equilibrium without any constraint.

(2) Let A be the given point, and B any other point in the plane of the forces: apply forces of constraint P and Q at A and B respectively as in (1). Then we shew as before that Q is zero. The forces *including* P must satisfy the necessary conditions of equilibrium: therefore the algebraical sum of their resolved parts in each of the two given directions is zero; but the algebraical sum of the resolved parts of the forces *excluding* P, in each of these two directions is zero, i.e. the resolved part of P in each of these directions is zero. But as the resolved part of a force is only zero, in a direction perpendicular to the force, P itself must be zero.

Hence the body is in equilibrium without constraint.

Case (3) can be proved in a similar way.

53. Prop. *Two equal forces acting in opposite directions along the same straight line on a rigid body, but not necessarily on the same particle, keep it in equilibrium.*

This is obvious as the two forces clearly satisfy the sufficient conditions of equilibrium given in the last article.

This proposition is essentially the same as the principle known as the *Transmissibility of Force*, which is generally assumed as an experimental fact, but which we prefer to deduce as above from the Laws of Motion. The formal statement of that principle is as follows: *when a force acts on a rigid body, it is indifferent on what particle in the line of action it acts, provided that particle is part*

STATICS OF SYSTEMS OF PARTICLES.   61

of the body, or rigidly connected with it. This follows directly from the proposition just proved. For let A, B be any two particles, in the line of action of the force P, and rigidly connected with the body. We have just proved that a force Q equal and opposite to P, will counterbalance it, provided Q acts at a point in AB rigidly connected with the body: hence the force P counteracts Q, whether P acts at A or at B. As regards its effect on the body as a whole, we may say then, that it is indifferent at which point we apply the force P. It is however in this sense only, that it is indifferent; if we take into consideration the internal forces brought into play in the two cases, they will probably be very different.

Imagine, for instance, a sphere resting on a smooth horizontal plane; a force of a certain magnitude, and in a certain direction will give the sphere the same change of velocity, whether the force take the shape of a push behind or a pull in front, yet the internal forces in the sphere will be different in the two cases, as in the first case the tendency of the external force is to compress the sphere, whereas it has the opposite tendency in the second case.

The proof of the converse principle, viz. that if it is indifferent at which of two points a force is applied, the line of action of the force must be the line joining them, is obvious from what has gone before.

Ex. 1. A square lamina $ABCD$ is acted upon by a force of 3 lbs. along AB, 2 lbs. along CB, 1 lb. along CD, 2 lbs. along AD, $\sqrt{2}$ lbs. along CA, and $\sqrt{2}$ lbs. along BD: prove that it is in equilibrium.

Ex. 2. A weightless rod AB, 10 feet long, has weights of 7 lbs. hung at each end, and one of 11 lbs. at its middle point: a string is attached to a point 2 feet from A and after passing over a smooth peg vertically above the point of the rod to which it is attached, supports a weight of 10 lbs.: another string attached to a point 3 feet from B supports in a similar way a weight of 15 lbs. Prove that the rod is in equilibrium.

Ex. 3. A rigid rod AB, 20 inches long, is acted upon by the following forces: 3 lbs. at A along BA, $\sqrt{3}$ lbs. at right angles to AB, at a point 5 inches from A, 6 lbs. at a point 5 inches from B, and making an angle of $60°$ with the part of the rod on the same side as A, and $4\sqrt{3}$ lbs.

at B making an angle of 30° with AB produced. Prove that there will be equilibrium, provided all the forces are in one plane, and the 3rd force acts on the opposite side of the rod to the 2nd and 4th.

Ex. 4. $ABCDEF$ is a regular hexagonal lamina: prove that it is kept in equilibrium by the following seven forces: 2 lbs. along AB, CD, DE, FA, and AD, 5 lbs. along CB and 3 lbs. along FE.

54.* Prop. *A rigid body under the action of any system of forces, is in equilibrium, provided the algebraical sum of their moments about each edge of any given tetrahedron be zero.*

Let $ABCD$ be the given tetrahedron, such that the

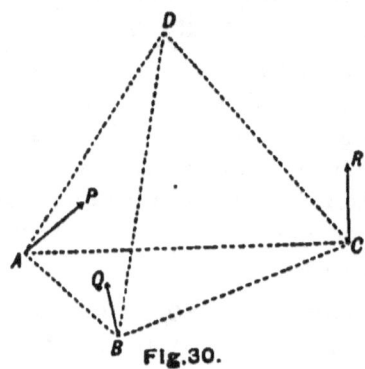

Fig. 30.

algebraical sum of the moments of the forces about each edge is zero. If the body is not in equilibrium under the action of the system of forces in question, it can be fixed (Arts. 50, 51) by applying suitable forces of constraint P, Q, R at A, B, and C respectively. Also Q may be taken perpendicular to AB, and R perpendicular to the plane ABC.

Since the body is in equilibrium under the action of the original forces *together with* P, Q and R, these forces must satisfy the necessary conditions of equilibrium. Therefore the algebraical sum of their moments about AB is zero; but the algebraical sum of the moments about AB of the original forces *alone* is zero, and the moments of both P and Q about AB are clearly zero, so that the moment of the remaining force R must be zero. R being perpendicular to the plane ABC, can neither intersect AB nor be parallel to it, so that its moment about AB can only vanish by R itself vanishing (Art. 38).

Similarly by taking moments about AC and AD, we see that the moment of Q about each of these lines is zero: hence Q must either be zero, or its line of action must lie in each of the planes BAC, BAD, i.e. must be the line AB; the latter alternative is out of the question, because Q is perpendicular to AB: Q must therefore be zero.

Again, by taking moments about BC, DB, and DC, we obtain that the moment of P about each of these lines is zero, i.e. that if P is not zero, its line of action lies in each of the planes BAC, BAD, DAC, which is impossible. P must therefore be zero. All the forces of constraint being zero, we see that the body is in equilibrium under the action of the original forces only.

55.* Prop. *A rigid body under the action of any system of forces is in equilibrium, provided the algebraical sum of their moments about each of any three given straight lines intersecting in a point, but not in one plane, be zero, and the algebraical sum of their resolved parts along each of these lines be zero also.*

Let OA, OB, OC be the straight lines, such that the algebraical sum of the moments of the forces about each of them is zero, and that of their resolved parts along each is zero also.

As in the last proposition, if the body is not in equilibrium, it may be fixed by applying suitable forces of constraint P, Q, R at O, A and B respectively; R may be taken perpendicular to the plane OAB, and Q perpendicular to the line OA. Then as in Art. 54, by taking moments about OA, R is found to be zero; and Q also by taking moments about OB and OC. But the algebraical

sum of the resolved parts of the original forces together
with P, along each of the lines OA, OB, OC must be zero;

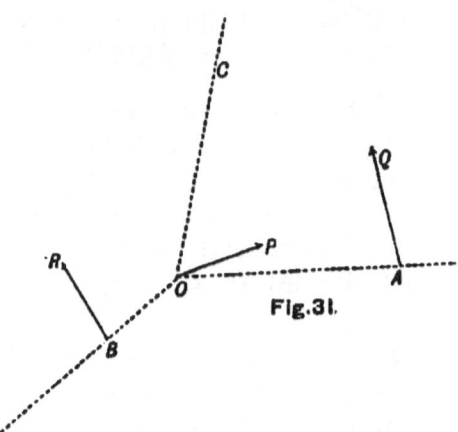

Fig. 31.

hence the resolved part of P along each of these lines
must also be zero, i.e. if P is not zero, it is perpendicular
to each of the lines OA, OB, OC, which is impossible as
they do not lie in one plane. P must therefore be zero.
The body is therefore in equilibrium under the action of
the original forces alone.

The *sufficient* conditions of equilibrium of any system
of forces acting on a rigid body can be expressed in many
ways, other than the two given above.

56. We have seen that if two systems of forces are
equivalent, either of them reversed will counteract the
other; hence it is sufficient for equivalence when both
systems are in the same plane, if any one of the following
sets of conditions holds. (1) If the algebraical sum of the
moments about each of three points in the plane but not
in the same straight line, of one system, be equal respectively to the corresponding sum of the other. (2) If the
algebraical sums of the moments about one point in the
plane, and of the resolved parts in two directions in it, of
one system be equal respectively to the corresponding
sums of the other. (3) If the algebraical sums of the

moments about each of two points in the plane, and of the resolved parts in one direction in the plane, not perpendicular to the line joining the two points, of one system, be equal respectively to the corresponding sums of the other.

Analogous conditions of equivalence can be obtained from Arts. 54, 55, for systems of forces which are not in one plane.

57. To find the resultant action on a body of a weightless string stretched round it.

Let $PABCDQ$ be a string stretched over a body, A and D being the points where the string leaves the body. The forces acting on the part

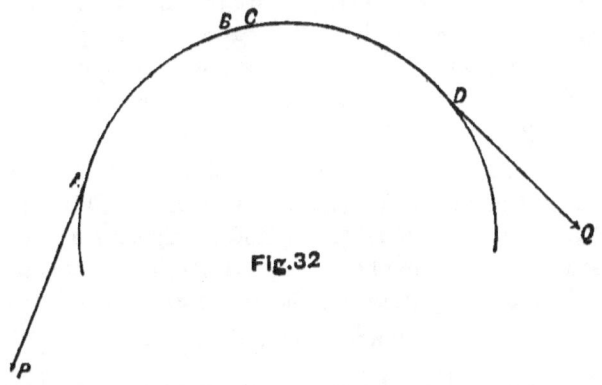

Fig. 32

$ABCD$ of the string are the force due to the part PA, or the tension at A along AP, the tension at D along DQ, and the innumerable actions of the body at every point of $ABCD$. Since this portion of the string is in equilibrium, the two tensions counteract all these actions along $ABCD$, i.e. they just balance the resultant of all these actions. But by Newton's Third Law, the resultant action of the string on the body is equal to, opposite to, and in the same straight line as, that of the body on the string. The two tensions counteract the latter of these resultants, i.e. they are equivalent to the former. We may therefore, in considering the equilibrium of the body, suppose that it is acted on *directly by the tensions at A and B*, instead of supposing, what is really the case, that these tensions act on the string $ABCD$, and so cause it to exert on the body the innumerable small forces, to which the tensions are equivalent.

We arrive at the same conclusion by regarding the body and the portion $ABCD$ of the string as one system of particles: in that case the tensions at A and B are forces *external* to the system, while the innumerable actions and reactions between the string and the body are *internal* forces.

Ex. 1. A smooth pulley is supported by a string which passes underneath it: find the weight of the pulley, if the tension of the string is 10 lbs. and the two parts not in contact with the pulley make angles of 30° with the vertical. *Ans.* $10\sqrt{3}$ lbs.

Ex. 2. Three smooth pegs are fastened in a vertical plane, so as to form an equilateral triangle whose base is horizontal and vertex downwards. A string with a weight 5 lbs. attached to each end, is passed under the lower peg and over the other two. Find the pressure on each peg.

Ans. $5\sqrt{3}$ lbs. on lower peg. $\frac{5}{2}(\sqrt{2}+\sqrt{6})$ lbs. on upper.

Ex. 3. A rope is passed several times round a fixed rough post, the tensions exerted at the ends of the two parts of the rope not in contact with the post, are 3 lbs. and $2\sqrt{2}$ lbs. respectively, and these two parts make an angle of 45° with one another. Find the resultant action of the rope on the post. *Ans.* $\sqrt{29}$ lbs.

Ex. 4. A circular cylinder (W) is placed with its axis horizontal on a smooth inclined plane: a weightless string is attached to a point in the plane and after passing over the cylinder supports a weight P, the straight portions of the string being respectively horizontal and vertical: shew that if there is equilibrium, the inclination of the plane to the horizon is
$$\tan^{-1}\{P/(P+W)\}.$$

58. We have seen how to obtain the resultant of two forces acting on the same particle; if now we have two forces acting on a rigid body, but not on the same particle, we can find a single force equivalent to them provided their lines of action either meet or are parallel, except in the case in which the forces are equal and opposite, but not in the same straight line. If their lines of action meet in a point, we may by the principle of the transmissibility of force, suppose each force to act at this point, and then their resultant is just what it would be if the forces really acted on a particle, situate there and rigidly connected with a body.

STATICS OF SYSTEMS OF PARTICLES. 67

59. *The Resultant of two parallel forces.* By Art. 56, a force in the same plane, whose resolved part in each of two directions in that plane equals the algebraical sum of the resolved parts of the two forces in the same direction, and whose moment about some point in the plane equals the algebraical sum of the moments of the two about that point, is the resultant.

Let A, B be two points where two parallel forces, P, Q respectively act. The first two conditions are satisfied

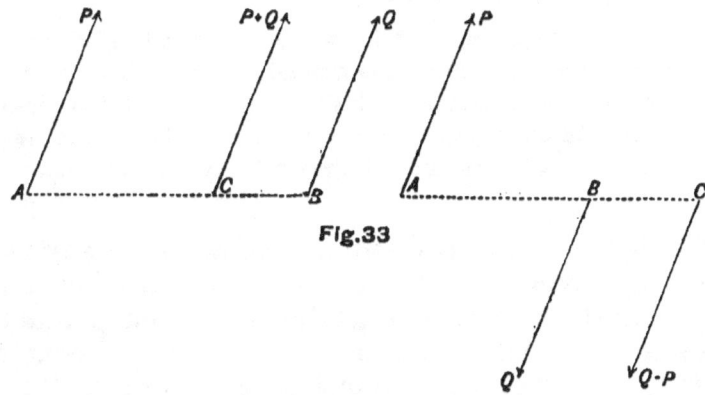

Fig. 33

by a force which acts in the same direction as P and Q, and is equal to their algebraical sum. The required force must then be parallel to the other two, and at such distances from them that their moments about any point in it are equal in magnitude but opposite in sign. It will therefore be between them if the signs of P and Q are the same, but not otherwise: its distances from them must be inversely proportional to their magnitudes. Hence if C be the point where its line of action meets AB, $P \cdot AC = Q \cdot BC$. When P and Q act in opposite directions, the greater force will clearly lie between the less and the resultant.

Cor. The position of C is independent of the direction of the forces, so long as they remain parallel.

If the forces P and Q are equal in magnitude and

opposite in sign, the preceding solution fails, and we can find no single force, whose effect is equal to that of the two together. Two such forces constitute a *couple*.

Ex. 1. Four forces, P, $2P$, $3P$, and $4P$ act along the sides taken in order of a square: find their resultant.

Ans. $2P\sqrt{2}$, acting parallel to the diagonal through the corner where $2P$, and $3P$ meet, and at a distance from it $\frac{5}{8}\sqrt{2}$ times a side of the square.

Ex. 2. A uniform beam 4 ft. long is supported in a horizontal position by two props which are 3 feet apart, so that the beam projects one foot beyond one of the props: shew that the pressure on one prop is double the pressure on the other.

Ex. 3. If a bicycle and its rider weigh 60 lbs. and 10 stone respectively, find how the pressure on the ground is divided between the two wheels, whose points of contact with the ground are 3 ft. 6 inches apart, while the points through which the weights of the bicycle and rider act, are distant horizontally 7 in. and 6 in. respectively from the centre of the driving wheel. *Ans.* 170 lbs. and 30 lbs.

60. Since a rigid body under the action of any system of coplanar forces, can be fixed by two forces of constraint acting in that plane at two arbitrarily chosen points in it, the system must be equivalent to the forces of constraint reversed: but two forces in one plane can be replaced by a single one, unless they form a couple. Hence

Any system of forces in one plane is equivalent to a single force or a couple.

61. Prop. *If three forces maintain equilibrium, their lines of action must be in one plane, and either all meet in one point or be all parallel.*

Let P, Q, R be the three forces, Aa, Bb, Cc, their respective lines of action.

Since the algebraical sum of the moments of a system of forces in equilibrium about any line is zero, that of the moments of P, Q, R about AB vanishes: but as P and Q both intersect this line, each of their moments about it is zero, hence that of R about it must also be zero, i.e. Cc meets AB or is parallel to it.

Similarly we can shew that Cc meets, or is parallel to Ab. Therefore Cc either lies in the plane ABb or passes through A. In the first case R and Q are in the same plane, in the second R and P. But if two of the forces are in one plane, the third must also be in it, as its moment about every line in the plane must be zero. Hence all three forces are in one plane.

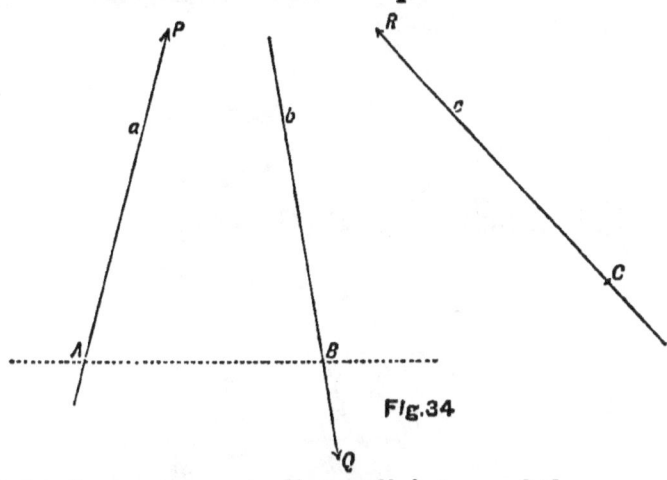

Fig. 34

If the forces are not all parallel, two of them meet and can be replaced by a single force, which is counterbalanced by the third force, and is therefore in the same straight line with it, i.e. the third force passes through the point of intersection of the other two.

Cor. Two forces, whose lines of action are not in one plane, cannot be equivalent to a single force.

62. *Def.* The *moment of a couple* is the algebraical sum of the moments of the two forces which form it, about any point in their plane.

This moment can easily be shewn to be independent of the position of the point and to be equal to the product of either force into the *arm*, i.e. the perpendicular distance between the lines of action of the forces.

For let P acting at A, and P acting in the opposite direction at B, form the couple. Then the algebraical sum of the moments of the two forces about O is

70 STATICS.

in Fig. (35), $\quad P(Oa + Ob) = P : ab,$

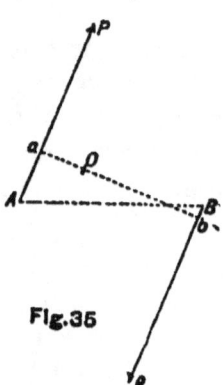

Fig.35

in Fig. (36), $\quad P(Oa - Ob) = P . ab,$

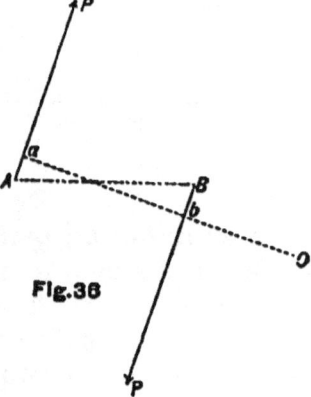

Fig.36

in Fig. (37), $\quad P(Ob - Oa) = P . ab,$

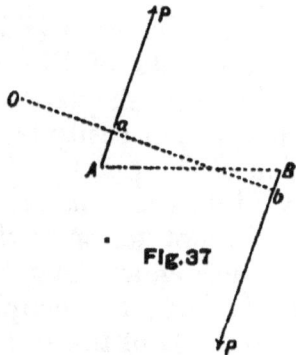

Fig.37

STATICS OF SYSTEMS OF PARTICLES.   71

where Oa, Ob are the perpendiculars from O on the lines of action of the forces.

If the body on which the couple acted were only free to turn round O, the tendency of the couple in all the above figures is to turn the body in the direction in which the hands of a watch move; the couples are said therefore to have moments of the same sign, or to be *like*; were the tendency of one of them to turn the body in the opposite direction, its moment would be of the opposite sign, and it would be *unlike* the other two.

63.   Prop. *Two like couples of equal moment, in the same or parallel planes, are equivalent to one another.*

(i)   When the couples are in the *same* plane.

In this case the two couples form two systems of forces in one plane, such that the algebraical sums of their moments about any point whatsoever in the plane are the same; therefore the systems are equivalent to one another (Art. 56).

(ii)   When the couples are in *parallel* planes.

Let P_1, P_2 be the two equal forces forming one of the couples, acting at the points A, B respectively.

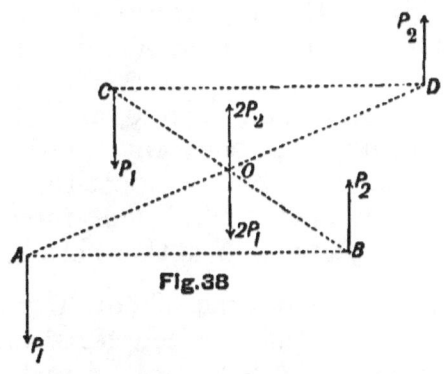

Fig. 38

In the plane in which the other couple acts, draw CD equal and parallel to AB. Join AD, BC, intersecting in O; then O bisects both AD and BC.

P_1 at A may be replaced by $2P_1$ at O and P_2 at D. P_2 at B may be replaced by $2P_2$ at O and P_1 at C: $2P_2$ and $2P_1$ at O counteract one another, so that we are left with P_1 at C and P_2 at D, as equivalent to the original couple. But these two forces constitute a couple *like* to the original one, equal to it and in the given plane parallel to it: therefore as the original couple is equal to *one* couple in the parallel plane, it is by (i) equal to *any* like couple of the same moment in that plane.

64.* The latter part of the last proposition might have been proved in a manner analogous to that adopted for the former, as follows.

Let A and B be the two couples: we shall prove that A and B reversed satisfy the sufficient conditions of equilibrium of Art. 55.

Take three straight lines, intersecting in a point, one perpendicular to the plane of each couple, and the other two in the plane of B.

It is obvious that the algebraical sum of the resolved parts of the four forces in each of these directions is zero: also the moments of A and B reversed, about the line perpendicular to their planes, are numerically equal but of opposite sign. Hence the algebraical sum of the moments of the four forces forming them about this line is zero. The moment of each of the forces forming B reversed about any line in their plane is zero, and the moments of the two forces forming A, about any line in the plane of B, are equal numerically but of opposite sign; the algebraical sum of the moments of all four forces about every straight line in the plane of B is therefore zero.

The six sufficient conditions of equilibrium of Art. 55 are therefore satisfied, and the couples A and B reversed balance one another; in other words A and B are equivalent.

STATICS OF SYSTEMS OF PARTICLES. 73

Ex. 1. Like parallel forces, each equal to P, act at three of the corners of a rhombus, perpendicular to its plane: at the other corner such a force acts that the four forces are equivalent to a couple: find the moment of the couple, provided the angle of the rhombus at which the last force acts is 60°. *Ans.* $2\sqrt{3}.Pa$, where a is a side of the rhombus.

Ex. 2. $ABCDEF$ is a regular hexagon: equal forces act along AB, BC, DE, EF, and two other forces, each double any one of the former forces, act along DC and AF: prove that they maintain equilibrium.

65.* Let us consider what we require to know to determine the effect of a couple on a rigid body. It is unnecessary to know the actual position of the plane in which the couple acts, but we must know the *direction* of the plane, i.e. the direction of a line to which it is perpendicular. We do not require to know the magnitude or direction of the forces which compose the couple, but we must know the *magnitude* of its *moment* and its *sign*, i.e. the direction in which it would tend to turn the body round a line perpendicular to its plane, the line being fixed and the body rigidly connected with it.

Now a straight line at right angles to the plane of the couple, and of length proportional to the magnitude of its moment, will represent the couple in the first two respects: also, if it be understood that the line is drawn in that direction in which the axis of a right-handed screw moves, when it rotates in the same way as the couple tends to turn the body, the *sign* of the couple will also be represented.

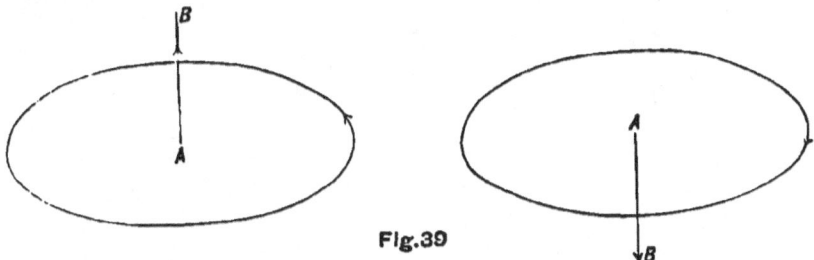

Fig.39

In fig. 39, if the arrowhead on the circle indicates the direction in which the couple would tend to turn the body about AB, supposing the latter fixed and the body rigidly connected with it, the sign of the couple in accordance with the above convention would be represented by *AB* and not by *BA*.

The line which thus completely represents the couple is termed the *axis* of the couple.

66.* We shall now prove that couples follow the *Parallelogram Law*, in other words, that

If from a point the axes representing two couples be drawn, and a parallelogram be constructed on these two axes as adjacent sides, the diagonal passing through the above-mentioned point is the axis of a couple equivalent to the two, i.e. of their resultant couple.

We may suppose the couples to consist of forces acting at the ends of a common arm, in which case the

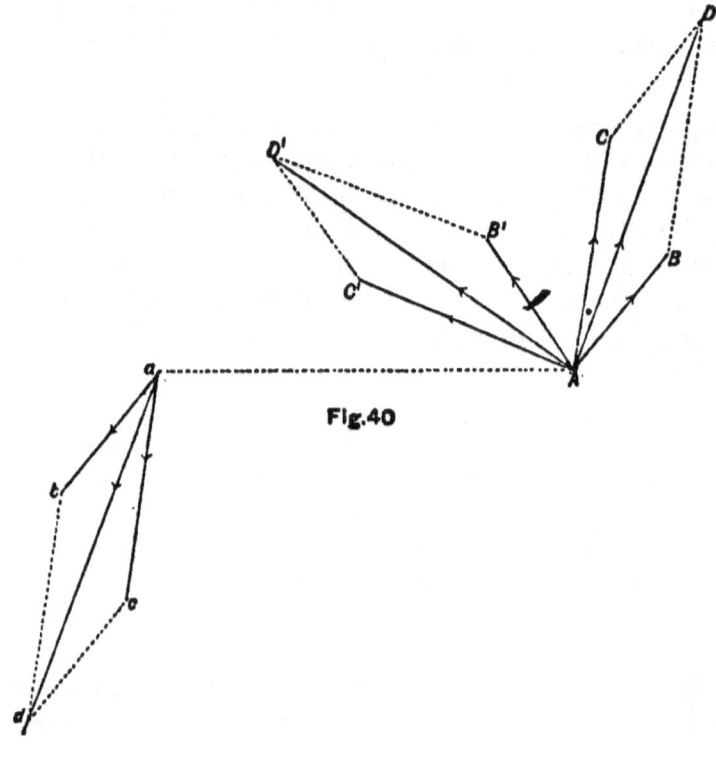

Fig. 40

STATICS OF SYSTEMS OF PARTICLES. 75

moments of the couples will be respectively proportional to the forces composing them.

Let Aa be the common arm, and let AB, ab represent the two equal and parallel forces forming the first couple, AC, ac those forming the second.

Draw AB' perpendicular to Aa and AB, equal to AB, and in the direction which by the convention of Art. 65 represents the sign of the first couple: similarly draw AC' perpendicular to Aa and AC, equal to AC and in the proper direction. Then AB' and AC' are the axes of the two couples.

Complete the three parallelograms, $ABCD$, $abcd$, $AB'C'D'$, and join AD, ad, AD'. These parallelograms are clearly equal in every respect, so that $AD = ad = AD'$. Also AD, ad are parallel, and AD' is perpendicular to AD.

But the two forces AB, AC are equivalent to AD, and the two ab, ac, to ad, so that the two couples are equivalent to AD, ad, which form a couple of which AD' is the axis. Hence the couples whose axes are AB', AC' are equivalent to a resultant couple of which AD' is the axis.

Cor. Hence we may deduce propositions relating to the composition and resolution of couples, analogous to those obtained in Arts. 19—26, 30—32, relating to the composition and resolution of forces.

67.* Prop. *Any system of forces acting on a rigid body can be reduced to a single force acting at any arbitrarily chosen point and a couple.*

Let A be the arbitrarily chosen point, P any one of the forces.

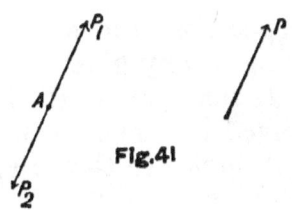

Fig. 41

We shall not alter the effect of the forces by applying at A two forces P_1, P_2 each equal and parallel to P, and in opposite directions to one another. P_2 which is opposite to P, forms with P a couple. Hence P is equivalent to P_1 at A, and a couple.

The couple vanishes in the case in which A lies in P's line of action.

Similarly we may replace each of the other forces by a force at A, equal to it and in the same direction, and a couple.

The whole system thus reduces to a series of forces at A, respectively equal to and in the same direction as the several original forces, and a series of couples. But the forces at A are equivalent to a single resultant at A, and the couples to a single resultant couple.

Cor. The magnitude and direction of the single resultant is the same wherever A is, and the resultant couple is the same for all positions of A in a line parallel to the single resultant force.

68.* Prop. *Any system of forces acting on a rigid body is equivalent to a single force and a couple whose axis is parallel to the direction of the single force.*

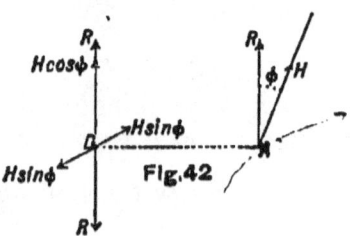

Fig. 42

By the last proposition, the system is equivalent to a single force R acting at any given point A, and a couple H. If the axis of H make an angle ϕ with the direction of R, it may be resolved into $H \cos \phi$ in the direction of R and $H \sin \phi$ at right angles to that direction.

STATICS OF SYSTEMS OF PARTICLES. 77

Draw AB perpendicular to both R and the axis of H, and make AB equal to $(H \sin \phi)/R$; then, applying at B two forces equal and parallel to R, but in opposite directions to one another, the system is equivalent to R at B in its original direction, the couples $H \cos \phi$, $H \sin \phi$, and the two forces R at A, and R in the opposite direction at B. But the last two forces are equivalent to a couple whose axis is at right angles to both R and AB, i.e. is in the same straight line as the axis of the couple $H \sin \phi$: its moment is $R \cdot AB$ or $H \sin \phi$. If AB be drawn as in fig. 43, the axes of these two couples are in opposite directions by the convention of Art. 65: the two couples therefore counteract one another, and we are left with R at B and the couple $H \cos \phi$ whose axis is along R's direction. Such a force and couple together form what is called a *wrench*.

R's line of action through B is termed *Poinsot's Central Axis* for the corresponding system of forces.

The algebraical sum of the moments of the system of forces about the axis of H through A is H, about a line through A, making an angle θ with AH, the sum of their moments is $H \cos \theta$. Hence AH is called the *axis of principal moment* at A, as the sum of the moments of the forces about it is greater than that about any other line through A.

69.* Prop. *The algebraical sum of the moments of the forces about Poinsot's Central Axis is less than that about any other axis of principal moment.*

For (Art. 68) (fig. 42) the sum of the moments about the central axis is $H \cos \phi$, whereas the sum of the moments about the axis of principal moment at A is H.

For this reason the Central Axis is sometimes termed the *Axis of Least Principal Moment*.

70.* Prop. *Any system of forces acting on a rigid body can be reduced to two equal forces equally inclined to the Central Axis.*

For let OG be the central axis, R being the single resultant force, and H the moment of the resultant couple whose axis is OG.

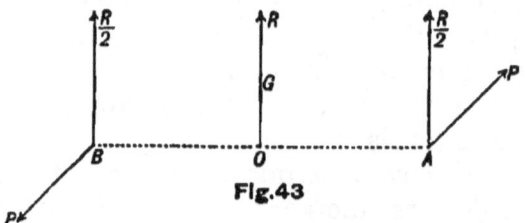

Fig. 43

Through O draw AOB perpendicular to OG, and make $OA = OB$.

We can replace R by $\frac{1}{2}R$ at A, and $\frac{1}{2}R$ at B, each in the same direction as R; we can replace the couple H by a force P at A, and a force P at B, each perpendicular to the plane GOA, but in opposite directions, provided $P = H/AB$.

The resultant of P and $\frac{1}{2}R$ at A, and that of P and $\frac{1}{2}R$ at B, will clearly be equal to one another and will make equal angles with $\frac{1}{2}R$, i.e. with the Central Axis.

71. *Recapitulation.* Regarding any body at rest whatsoever, as a collection of particles each of which is at rest, we can assert that the algebraical sum of the resolved parts in any direction, of all the forces, internal as well as external, acting on the body is zero: also that the algebraical sum of their moments about any line is zero. But as by Newton's Third Law the internal forces consist entirely of pairs, which are equal, opposite and in the same straight line as one another, the algebraical sums of the resolved parts and of the moments of the internal forces are both zero. Any system of external forces which, together with internal ones, maintain a body in equilibrium, must therefore be such that the algebraical sum of their resolved parts in any direction is zero and that of their moments about any line is zero also.

Next, considering rigid bodies only, we shew that a body under the action of any system of external forces whatsoever can be fixed by the application of suitable forces at three arbitrarily chosen points, and that the direction of one of these forces may be taken perpendicular to the plane containing the three points, and that of another perpendicular to the line joining its point of application to the third point. When the forces are coplanar, the body can be fixed by applying suitable forces at any two points in the plane of the forces, the directions of both forces being in the plane and that of one perpendicular to the line joining the two points. From this proposition follow the *sufficient* conditions of equilibrium of a system of coplanar forces acting on a rigid body. These conditions may be given in three different forms, and each form is expressed algebraically by three equations. When the forces are not in one plane the sufficient conditions of equilibrium can be put in many different forms, and each form requires for its algebraical expression six equations.

Defining two forces as equivalent, when either counteracts the other reversed, we deduce the principle known as the 'Transmissibility of Force.'

The resultant of two parallel forces is obtained by finding from the sufficient conditions of equilibrium, a force which will counteract them, and then reversing it.

It is then shewn that if three forces maintain equilibrium, they must be coplanar and either concurrent or parallel.

Then we shew that two couples are equivalent when their moments are equal and their planes coincident or parallel; hence that couples can be represented by straight lines, and that they can be compounded and resolved like forces by the Parallelogram Law. It was shewn that any system of forces in one plane is equivalent to a single force or a couple, it can now be shewn that any system of forces whatsoever, acting on a rigid body, is equivalent to a single force and a couple acting in a plane perpendicular to the force, or to two equal forces, equally inclined to the Central Axis.

ILLUSTRATIVE EXAMPLES.

Ex. 1. If four forces acting along the sides of a quadrilateral are in equilibrium, prove that the quadrilateral is a plane one, and also, that if the quadrilateral can be inscribed in a circle, each force must be proportional to the length of the opposite side.

Let $ABCD$ be the quadrilateral. The forces along AB, BC have a resultant through B and in the plane ABC; similarly those along AD, DC

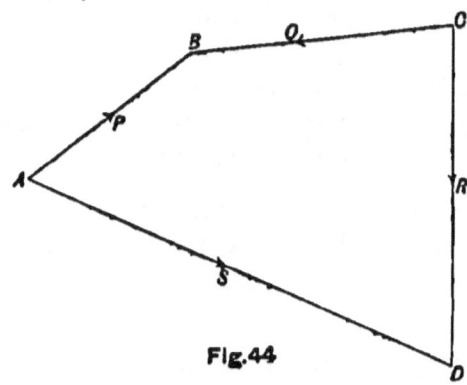

Fig.44

have a resultant through D and in the plane ADC. But as the four forces are in equilibrium, these two resultants must be in the same straight line, BD, i.e. BD is in each of the planes ABC, ADC and the quadrilateral is a plane one.

When $ABCD$ can be inscribed in a circle let P, Q, R, S be the forces along AB, CB, CD, AD, respectively.

Since the forces are in equilibrium, the algebraical sum of their moments about A is zero:

$$\therefore Q \cdot AB \sin B - R \cdot AD \sin D = 0;$$

$$\therefore \frac{Q}{AD} = \frac{R}{AB} = \text{similarly, } \frac{P}{CD} = \frac{S}{BC}.$$

Ex. 2. $ABCD$ is a quadrilateral, and two points P, Q are taken in AD, BC such that $AP : PD = CQ : QB$. From P, Q, straight lines PP', QQ' are drawn parallel to, equal to, and in the same directions as BC and DA respectively. Shew that forces represented by AB, CD, PP', QQ' are in equilibrium.

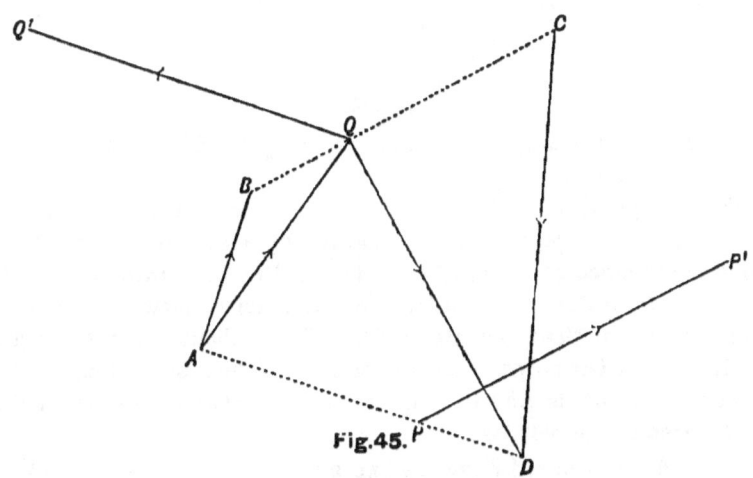

Fig. 45.

The force PP' can be replaced by two forces parallel to it, at A and D: the force at $A : PP' = PD : AD = BQ : BC$;

$$\therefore \text{force at } A = BQ;$$

similarly, force at $D = QC$.

The two forces AB, BQ, acting at A, are, by the triangle of forces, equivalent to AQ; and the two QC, CD, at D, to QD. Hence the four original forces are equivalent to AQ, QD, and QQ', all acting through Q, and represented by the sides of the triangle AQD, taken in order. They are therefore in equilibrium.

Ex. 3. A system of forces represented by the sides of a plane polygon, taken in order, is equivalent to a couple, whose moment is represented by twice the area of the polygon.

Let the forces be represented by the sides AB, BC, CD, DE, EF, FA, of the polygon $ABCDEF$.

We know that if the forces are not in equilibrium, they are equivalent to a single resultant or a couple (Art. 60).

But as the algebraical sum of their resolved parts in any direction is

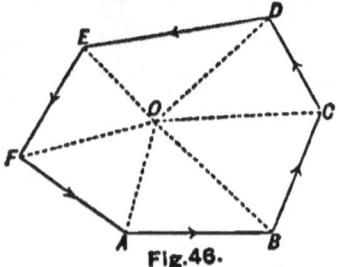
Fig. 46.

zero, their resultant is zero, i.e. they are in equilibrium, if they are not equivalent to a couple.

Take any point O, and join OA, OB, OC, &c.: then the moment of AB about O is measured numerically by twice the area of the triangle OAB, since the area of OAB is equal to $\frac{1}{2}AB$ into the perpendicular from O on AB: and similarly for the other moments. Hence the algebraical sum of the moments of AB, BC, &c. about O is measured by twice the area of the polygon, i.e. is not zero. The system then must be equivalent to a couple, and the moment of this couple is represented by twice the area of the polygon.

Ex. 4. A uniform rod hangs by two strings of lengths l, l', fastened to its ends and to two points in the same horizontal line, distant a

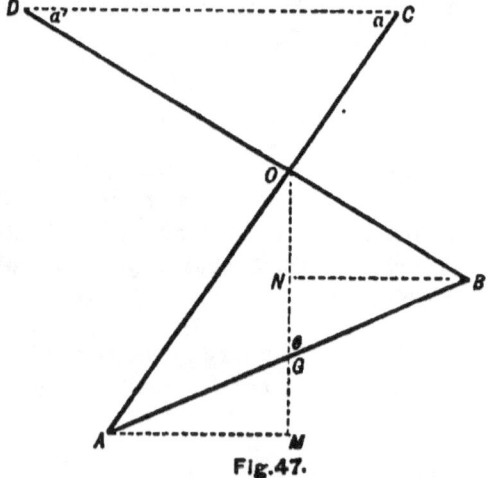
Fig. 47.

apart, the strings crossing one another. Find the position of equilibrium, and shew that if a, a' be the angles that l, l' make with the horizontal
$$\sin(a+a')(l'\cos a' - l\cos a) = a \sin(a-a').$$

Let θ be the angle which the rod AB makes with the vertical: let O be the point where the strings cross one another. Since the rod is in equilibrium under the action of three forces, two of which, the tensions of the strings, meet in O, the third, the weight of the rod, passes through O. But the weight acts vertically through the middle point of the rod, which point G, must therefore be in a vertical line with O: hence the perpendiculars AM, BN on OG must be equal.

$$\therefore (l - OC) \cos \alpha = (l' - OD) \cos \alpha'.$$

But
$$\frac{OC}{\sin \alpha'} = \frac{OD}{\sin \alpha} = \frac{a}{\sin (\alpha + \alpha')},$$

$$\therefore \left\{ l - \frac{a \sin \alpha'}{\sin (\alpha + \alpha')} \right\} \cos \alpha = \left\{ l' - \frac{a \sin \alpha}{\sin (\alpha + \alpha')} \right\} \cos \alpha',$$

or
$$\sin (\alpha + \alpha') (l \cos \alpha - l' \cos \alpha') = a \sin (\alpha' - \alpha) \dots \dots (1).$$

If b be the length of AB, since the algebraical sums of the vertical and horizontal projections of AB, BD, DC, CA are both zero,

$$l \sin \alpha - b \sin \theta - l' \sin \alpha' = 0,$$
$$l \cos \alpha - b \cos \theta + l' \cos \alpha' - a = 0.$$

These equations with (1) enable us to obtain α, α', and θ, which determine the position of equilibrium.

The above is an example of a *geometrico-statical* problem, in which the position of equilibrium, which must clearly exist, is required, and is obtained from geometrical considerations.

If the weight of the rod be given, the other unknown quantities, the tensions of the strings, can be obtained by using two more conditions of equilibrium, since there are three, and one only has been used. As there are five unknown quantities, and only three sufficient conditions of equilibrium, we must have two geometrical conditions in order to completely solve the problem.

Ex. 5. A uniform heavy rod of length $2a$ is placed across a smooth horizontal rail and rests with one end against a smooth vertical wall, the distance of which from the rail is h: shew that the angle the rod makes with the horizon is $\cos^{-1} (h/a)^{\frac{1}{3}}$.

Let θ be the inclination of the rod to the horizon, in the position of equilibrium. The forces acting on the rod AB are its weight vertically downwards through G, its middle point, the reaction of the wall, horizontally through A, and that of the rail C, at right angles to AB. These

forces must therefore meet in a point D. Since ADG, ACD are right angles,
$$AD^2 = AC \cdot AG,$$
$$\therefore a^2 \cos^2 \theta = ah \sec \theta,$$
$$\therefore \cos^3 \theta = h/a.$$

Or, we might have proceeded thus: let R be the reaction of the wall, S that of the peg, and W the weight of the rod. Resolving vertically, we have
$$W - S \cos \theta = 0 \dots\dots\dots\dots\dots\dots\dots\dots\dots(1).$$

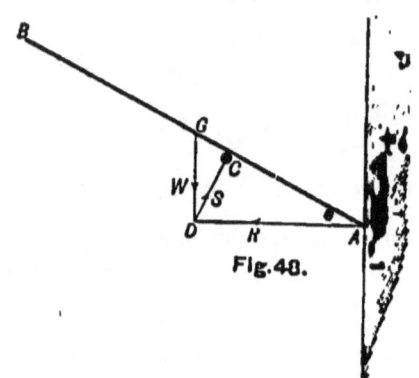

Fig. 48.

Taking moments about A,
$$S \cdot AC = W \cdot AD,$$
$$\therefore Sh \sec \theta = Wa \cos \theta \dots\dots\dots\dots\dots\dots(2).$$
From (1) and (2) $\cos^3 \theta = h/a.$

Resolving horizontally,
$$R - S \sin \theta = 0 \dots\dots\dots\dots\dots\dots\dots(3).$$
Hence R and S can be obtained.

The advantage of resolving vertically and taking moments about A is that in neither case does the force R come into the corresponding equation.

Ex. 6. Shew that the greatest inclination to the horizon at which a uniform rod can rest, partly within and partly without a fixed smooth hemispherical bowl, is $\sin^{-1} (\frac{1}{3}\sqrt{3})$.

Let $ADEC$ be the circular section of the complete sphere, made by the vertical plane containing the rod AB, which rests against the edge of the bowl at C. COD is the horizontal diameter of the sphere through C.

The rod is kept in equilibrium by its weight through G, its middle point, the reaction of the bowl at A, along the normal AO, and that at C perpendicular to AB, and therefore meeting AO on the sphere at E.

Since these three forces pass through one point, GE must be a vertical line.

Let $ACO = \theta$, r = radius of the bowl.

Then
$$AC = AE \cos EAC = 2r \cos \theta \quad \ldots\ldots\ldots\ldots \quad (1),$$

$$AG = AE \cdot \frac{\sin AEG}{\sin EGA} = 2r \frac{\cos 2\theta}{\cos \theta} \quad \ldots\ldots\ldots\ldots \quad (2),$$

since
$$AEG = \tfrac{1}{2}\pi - EOF = \tfrac{1}{2}\pi - 2\theta.$$

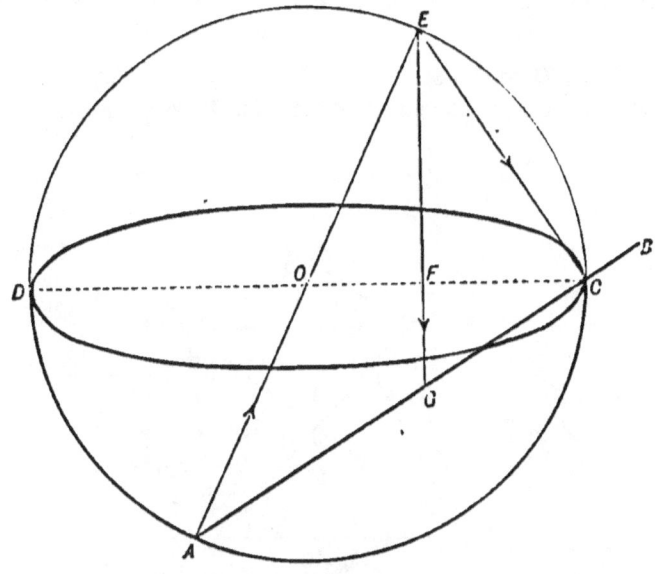

Fig. 49.

Since AG is half the rod, (2) determines the position of equilibrium.

Let
$$m = \frac{AG}{AC},$$

$$\therefore m = \frac{\cos 2\theta}{\cos^2 \theta} = \frac{2\cos^2 \theta - 1}{\cos^2 \theta},$$

$$\therefore \cos^2 \theta = \frac{1}{2 - m}.$$

θ clearly has its greatest value when m has its least value, i.e. when $m = \tfrac{1}{2}$, since AG cannot be less than half AC.

Hence the greatest value of θ is given by
$$\cos^2 \theta = \frac{1}{2-\frac{1}{2}} = \frac{2}{3},$$
or
$$\sin \theta = \frac{1}{\sqrt{3}}.$$

Ex. 7. Four equal spheres rest in contact at the bottom of a smooth spherical bowl, their centres being in a horizontal plane. Shew that, if another equal sphere be placed upon them, the lower spheres will separate if the radius of the bowl be greater than $(2\sqrt{13}+1)$ times the radius of a sphere.

Let A, B, C, D be the centres of the four spheres respectively, O that of the upper sphere, O' that of the spherical bowl. Then $AB, BC, CD,$

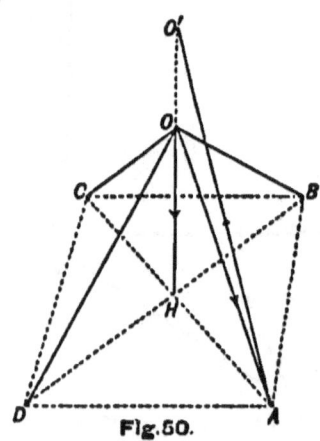

Fig. 50.

DA, OA, OB, OC, OD are each equal to the diameter $(2r)$ of any one of the spheres. O and O' are clearly in the vertical line through H, the intersection of the diagonals of the square $ABCD$.

Then
$$\cos OAH = \frac{AH}{OA} = \frac{\frac{1}{\sqrt{2}}AB}{OA} = \frac{1}{\sqrt{2}},$$
$$\therefore \angle OAH = 45°.$$

When the lower spheres are just on the point of separating, there is no pressure between any two of them, so that each of them is in equilibrium under the action of its weight, the pressure of the upper sphere, and that of the hollow sphere. Let W be the weight of each sphere, R

STATICS OF SYSTEMS OF PARTICLES. 87

the reaction between the upper and any of the lower spheres. From the equilibrium of the upper sphere, resolving vertically,

$$W - 4R \cdot \frac{1}{\sqrt{2}} = 0,$$

$$\therefore R = \frac{W}{2\sqrt{2}}.$$

The resultant of W acting vertically, and $\frac{W}{2\sqrt{2}}$ along OA, on the sphere whose centre is A, makes with the vertical the angle $\tan^{-1} \dfrac{\frac{W}{2\sqrt{2}} \cdot \frac{1}{\sqrt{2}}}{W + \frac{W}{2\sqrt{2}} \cdot \frac{1}{\sqrt{2}}}$,

i.e. $\tan^{-1} \frac{1}{5}$.

But this resultant is equal and opposite to the pressure of the bowl which acts along AO'.

Therefore $\tan AO'H = \frac{1}{5}$,

$$\therefore \frac{AH}{O'A} = \sin AO'H = \frac{\frac{1}{5}}{\sqrt{(1+\frac{1}{25})}} = \frac{1}{\sqrt{26}}.$$

$$\therefore O'A = \sqrt{26} \cdot AH = 2r\sqrt{13}.$$

But the radius of the hollow sphere is equal to $O'A$ together with r, therefore radius of the bowl $= (2\sqrt{13} + 1)r$.

If the bowl is any larger, O' will be further from H, and for the pressure of the bowl to counteract the resultant of the other forces on the sphere (centre A), we shall have to suppose that the actions of the two adjacent lower spheres on it are *towards* their respective centres instead of *away from* them. But as the spheres are incapable of exerting such forces, equilibrium is not possible, i.e. the spheres will separate.

Ex. 8. A heavy bar, AB, is suspended by two equal strings of length l, which are originally parallel: find the couple which must be applied to

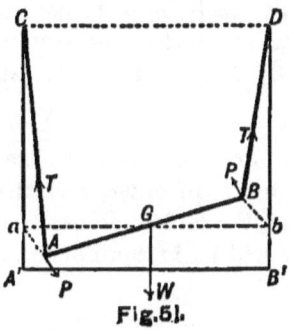

Fig. 51.

the bar to keep it at rest after it has been twisted through an angle θ in a horizontal plane.

Let C, D be the fixed ends of the strings; CA', DB' the original vertical positions of the strings.

Draw Aa, Bb at right angles to CA', DB' respectively. Join ab cutting AB in its middle point G. Let $2a$ be the length of AB, and $\phi =$ angle aCA or bDB.

Then
$$CA \sin aCA = aA = 2AG \sin \tfrac{1}{2}aGA,$$
$$\therefore l \sin \phi = 2a \sin \tfrac{1}{2}\theta \dots\dots\dots\dots\dots\dots(1).$$

Let T be the tension of either string: they will from symmetry be the same.

Let P be the magnitude of the force which applied horizontally in opposite directions at A and B, at right angles to AB, will keep the rod in equilibrium.

Resolving vertically, we have
$$W - 2T \cos \phi = 0.$$

Taking moments about the line of action of W, we have
$$2P \cdot a - 2T \sin \phi \cdot a \cos \tfrac{1}{2}\theta = 0.$$

Hence
$$2Pa = \frac{aW \sin \phi \cos \tfrac{1}{2}\theta}{\cos \phi} \dots\dots\dots\dots\dots(2).$$

(1) and (2) enable us to determine $2Pa$ in terms of a, l, W and θ.

In this example we have assumed as obvious that a couple only is required to maintain equilibrium: it can be shewn however that the values we have obtained for P and T will satisfy the six conditions of equilibrium of Art. 55.

EXAMPLES.

1. Four points A, B, C, D lie on a circle and forces act along the chords AB, BC, CD, DA, each force being inversely proportional to the corresponding chord: prove that the resultant passes through the common points of (1) AD, BC; (2) AB, DC; (3) tangents at B, D, and (4) tangents at A and C.

2. If six forces acting on a body be completely represented, three by the sides of a triangle taken in order, and three by the sides of the triangle formed by joining the middle points of the sides of the original triangle, prove that they will be in equilibrium if the parallel forces act

in the same direction, and the scale on which the first three forces are represented be four times as large as that on which the last three are represented.

3. Forces P, Q, R act along the sides of a triangle ABC, and their resultant passes through the centres of the inscribed and circumscribed circles: prove that
$$\frac{P}{\cos B - \cos C} = \frac{Q}{\cos C - \cos A} = \frac{R}{\cos A - \cos B}.$$

4. Prove that a uniform rod cannot rest entirely within a smooth hemispherical bowl, except in a horizontal position.

5. If a uniform heavy rod be supported by a string fastened at its ends, and passing over a smooth peg; prove that it can only rest in a horizontal or vertical position.

6. A heavy equilateral triangle hung upon a smooth peg by a string, the ends of which are attached to two of its angular points, rests with one of its sides vertical; show that the length of the string is double the altitude of the triangle.

7. A fine string $ACBD$ tied to the end A of a uniform rod AB of weight W, passes through a fixed ring at C, and also through a ring at the end B of the rod, the free end of the string supporting a weight P; if the system be in equilibrium prove that $AC : BC :: 2P + W : W$.

8. A horizontal rod, the ends of which are on two inclined planes, is in equilibrium: if α, β be the inclinations of the planes, prove that the centre of gravity of the rod divides it into two parts in the ratio of $\tan \alpha$ to $\tan \beta$.

9. A uniform heavy rod AB has the end A in contact with a smooth vertical wall, and one end of a string is fastened to the rod at a point C such that $AC = \frac{1}{4}AB$, and the other end of the string is fastened to the wall; find the length of the string if the rod is in equilibrium in a position inclined to the vertical.

10. A cylindrical ruler whose radius is a, and length $2h$ rests on a horizontal rail with one end pressing against a smooth vertical wall, to which the rail is parallel. Shew that the angle the axis of the ruler makes with the vertical is given by $(h \sin \theta + a \cos \theta) \sin^2 \theta + 2a \cos \theta = b$, where b is the distance of the rail from the wall.

11. Two equal heavy spheres of one inch radius are in equilibrium within a smooth spherical cup of three inches radius. Show that the

pressure between the cup and one of the spheres is double the pressure between the two spheres.

12. Along each side taken in order of a polygon inscribed in a circle, acts a force whose magnitude is proportional to the sum of the lengths of the two adjacent sides: prove that the system of forces is equivalent to a system of forces acting along the tangents at the corners of the polygon, each such force being proportional to the length of the chord joining the two adjacent points.

13. $ABCD$ is a quadrilateral: forces act along the sides AB, BC, CD, DA measured by $\alpha, \beta, \gamma, \delta$ times those sides respectively. Shew that if there is equilibrium
$$\alpha\gamma = \beta\delta.$$

Shew also that $\triangle ABD / \triangle ABC = \alpha(\gamma - \beta)/\delta(\beta - \alpha)$.

14. Into the top of a fixed smooth sphere of radius a is fitted firmly a fine smooth vertical rod. A bar of length $2b$ has at one end a ring which slides on the rod; and the bar rests on the sphere. Shew that in equilibrium the angle (α) the bar makes with the horizontal is given by
$$a \sin \alpha = b \cos^3 \alpha.$$

15. Forces P, Q, R act along the sides BC, CA, AB of a triangle; shew that their resultant will act along the line joining the centre of the circumscribing circle with the orthocentre if
$$P : Q : R :: \frac{\cos B}{\cos C} - \frac{\cos C}{\cos B} : \frac{\cos C}{\cos A} - \frac{\cos A}{\cos C} : \frac{\cos A}{\cos B} - \frac{\cos B}{\cos A}.$$

16. A kite (weight P) having a tail (weight Q) is stationary, with a normal to its face, the direction of the wind, which is horizontal, and the string in the same vertical plane. The tail is attached at a distance a below the kite's centre of gravity, the string at a distance b above. Shew that, neglecting the action of the wind on the tail, the inclination of the kite to the horizon is given by the equation
$$\Pi b \sin^2 \theta = \{Pb + Q(a+b)\} \cos \theta,$$
where Π is the pressure on the kite, when placed perpendicular to the wind's direction.

17. Forces act at the middle points of the sides of a rigid polygon in the plane of the polygon; the forces act at right angles to the sides, and are respectively proportional to the sides in magnitude: shew that the forces will be in equilibrium if they all act inwards or all act outwards.

18. Show that it is impossible to arrange six forces along the edges of a tetrahedron so as to form a system in equilibrium.

19. On the sides of a right-angled triangle ABC squares are described, the square $BCDE$ on the hypotenuse on the same side of BC as A, and the squares $CAFG$, $ABHK$ on CA, AB on the opposite side of each to the triangle: prove that the forces represented by the straight lines AB, BC, CA, BH, HK, KA, CD, DE, EB, AF, FG, GC will form a system in equilibrium.

20. If four parallel forces balance each other, let their lines of action be intersected by a plane, and let the four points of intersection be joined by six straight lines so as to form four triangles; then prove that each force is proportional to the area of the triangle whose angles are in the lines of action of the other three.

21. Two rings of weight P and Q respectively, slide on a string, whose ends are fastened to the extremities of a straight rod inclined at an angle θ to the horizon: on the rod slides a light ring through which the string passes so that the heavy rings are on different sides of the light ring. Prove that in the position of equilibrium the inclination ϕ of those parts of the string next the weightless ring, to the rod, is given by the equation $\tan\phi/\tan\theta = (P+Q)/(P \sim Q)$.

22. An elastic string passes round three equal right-circular cylinders so that when each cylinder touches the other two along a generating line, the string is just not stretched: shew that if the system be placed on a smooth horizontal plane, the inclination (θ) of the plane containing the axis of the upper cylinder, and that of either of the lower ones to the horizontal, in the position of equilibrium, is given by the equation $(\pi+3)W = 2\lambda(2\cos\theta - 1)\tan\theta$. ($W$ is the weight of the upper cylinder, and λ is the modulus of elasticity.)

23. Two equal circular discs, of radius r, with smooth edges are placed on their flat sides in the corner between two smooth vertical planes inclined at an angle 2α and touch each other in the line bisecting the angle; the radius of the least disc which may be pressed between them without causing them to separate $= r(1-\cos\alpha)/\cos\alpha$.

24. A rectangular lamina $ABCD$ is supported with its plane vertical and one edge AB in contact with a smooth vertical wall, by an endless string which passes through smooth rings, one fixed to the wall at A, and two others P, Q fixed in the sides AB, CD of the lamina respectively

so that PQ is parallel to AD. Prove that the string has the least tension consistent with equilibrium when the position of Q is such that

$$BC/2AB = \tan \tfrac{1}{2} AQD.$$

25. Forces act through the angular points of a tetrahedron perpendicular to the opposite faces and proportional to them. Prove that they are in equilibrium if they all act either inwards or outwards.

26. AC, BD are two non-intersecting straight lines of constant length; prove that the effect of forces represented in every respect by AB, BC, CD, DA is the same, so long as AC, BD remain parallel to the same plane, and their projections on that plane are inclined at a constant angle to one another.

27. A flat semi-circular board with its plane vertical and curved edge upwards rests on a smooth horizontal plane, and is pressed at two given points of its circumference by two beams which slide in smooth vertical tubes: find the ratio of the weights of the beams to one another when the board is in equilibrium.

28. An endless string is placed round two equal cylinders and the system is suspended from a peg so that the line joining the centres of the cylinders is horizontal. If the pressure between the cylinders be equal to twice the weight of either of them; prove that the length of the string : the radius of either cylinder :: $4(2 + \tan^{-1} 2) : 1$.

29. A homogeneous circular cylinder rests on two smooth planes inclined to the horizon at angles α and β in opposite directions, so that its axis is at right angles to the line of intersection of the planes. Prove that the inclination θ of the base to the vertical in the position of equilibrium is given by

$$\tan \theta = \frac{a \sin(\alpha - \beta)}{r \sin(\alpha + \beta) + 2a \sin \alpha \sin \beta},$$

where r is the radius of the base and $2a$ the length of the cylinder.

30. In a triangular lamina ABC, AD, BE, CF are the perpendiculars on the sides, and forces represented by the lines BD, CD, CE, AE, AF, BF are applied to the lamina; prove that their resultant will pass through the centre of the circle described about the triangle.

31. An elliptic lamina rests against an inclined plane (α) being supported by a string attached to the extremity of its minor axis, so that its major axis is vertical and the plane of the ellipse is perpendicular to

the inclined plane. Shew that the inclination of the string to the vertical is $\tan^{-1} b \sqrt{(a^2 + b^2 \tan^2 a)/(a^2 - b^2)}$.

32. A uniform bar of length a rests suspended by two strings of lengths l and l' fastened to the ends of the bar and to two fixed points in the same horizontal line at a distance c apart. If the directions of the strings, being produced, meet at right angles, prove that the ratio of their tensions is $al + cl' : al' + cl$.

33. Two weights P, P are attached to the ends of two strings which pass over the same smooth peg and have their other extremities attached to the ends of a beam AB, the weight of which is W; shew that if θ be the inclination of the beam to the horizon $(a + b) \tan \theta = (a - b) \tan a$; a, b being the distances of the centre of gravity of the beam from its ends, and $\sin a = W/2P$.

34. A string 9 feet long has one end attached to the extremity of a smooth uniform heavy rod two feet in length, and at the other end carries a ring which slides upon the rod. The rod is suspended by means of the string from a smooth peg; prove that if θ be the angle which the rod makes with the horizon, then $\tan \theta = 3^{-\frac{1}{3}} - 3^{-\frac{2}{3}}$.

35. A triangle formed of three smooth rods is fixed horizontally, and a homogeneous sphere rests on it. Prove that the pressure on each rod is proportional to its length.

36. A sphere rests on three smooth pegs, which lie in a horizontal plane, and are at distances a, b, c from one another, prove that the pressures on the pegs are in the ratios
$$a^2 (b^2 + c^2 - a^2) : b^2 (c^2 + a^2 - b^2) : c^2 (a^2 + b^2 - c^2).$$

37. ABC, $A'B'C'$ are two triangles inscribed in the same circle; and forces proportional to the sides of the triangle act along them, but in opposite directions round the two triangles. Prove that, if a, β, γ be the angles subtended at the centre of the circle by the sides of the one triangle, and a', β', γ' those subtended by the sides of the other, the forces will be in equilibrium if $\sin \tfrac{1}{2} a \sin \tfrac{1}{2} \beta \sin \tfrac{1}{2} \gamma = \sin \tfrac{1}{2} a' \sin \tfrac{1}{2} \beta' \sin \tfrac{1}{2} \gamma'$.

38. A, B, C, D are four points in space: four forces represented by $AB, AD, CB,$ and CD act along these lines: prove that they have a single resultant, the line of action of which is perpendicular to the shortest distance between the lines AB, DC, and also to that between AD, BC.

39. Three equal spheres are placed in contact on a smooth horizontal table, and a fourth equal sphere is placed upon them, and then a cone of semi-vertical angle a is placed over the pile of spheres. Prove that the cone will be lifted if its weight is less than $\dfrac{1}{\sqrt{2}} \tan a$ of the weight of a sphere.

40. A cylindrical shell, without a bottom, stands on a horizontal plane, and two smooth spheres are placed within it, whose diameters are each less whilst their sum is greater than that of the interior surface of the shell: shew that the cylinder will not upset if the ratio of its weight to the weight of the upper sphere be greater than $2c - a - b : c$, where a, b, c are the radii of the spheres and cylinder.

41. Three spheres of radius c are placed on a smooth horizontal table so that their points of contact with it are at the angular points of an equilateral triangle. A fourth sphere of radius a and weight W touches the table and each of the other spheres. An elastic string of natural length $2\pi c$ and modulus of elasticity μ is placed symmetrically round the first three spheres. If the fourth sphere is just on the point of ascending, shew that $2\pi c W = 27 \mu (a - c)$.

42. A uniform rod, length c and weight W is suspended from a fixed point by two equal elastic strings, the natural length of each being c and the modulus w. A particle of weight W is placed on the rod at a distance x from its middle point, and when the system is in equilibrium the rod makes an angle a with the vertical. If θ, ϕ are the angles the strings make with the vertical, prove that

$$\frac{x}{c} = \frac{\sin(\theta \sim \phi) - 2 \cot a \sin \theta \sin \phi}{\sin(\theta + \phi)} = \frac{\sin \theta \sim \sin \phi}{\sin a},$$

and obtain another equation connecting θ and ϕ.

43. A lamina in the form of an isosceles triangle of vertical angle a rests with its plane vertical and its two equal sides each in contact with a smooth peg, the pegs being in a horizontal line distant c apart: prove that the axis of the triangle is vertical or makes with it the angle $\cos^{-1}(h \sin a/3c)$. h is the length of the axis of the triangle.

44. Two strings of the same length have each of their ends fixed at each of two points in the same horizontal plane. A smooth sphere of radius r and weight W is supported upon them at the same distance

from each of the given points. If the plane in which either string lies makes an angle a with the horizon, prove that the tension of each $= Wa/8r \sin a$; a being the distance between the points.

45. A smooth semi-circular tube is just filled with $2n$ equal smooth beads that just fit the tube, and the whole is at rest in a vertical plane with the bounding diameter highest. If R_m be the pressure between the mth and $(m+1)$th beads from the top, then

$$R_m = W . \sin \frac{m\pi}{2n} \Big/ \sin \frac{\pi}{2n},$$

where W is the weight of a bead.

Hence deduce that when the beads are diminished indefinitely in size, the pressure between any two is proportional to their depth below the top one.

46. A smooth rod passes through a smooth ring at the focus of an ellipse whose major axis is horizontal and rests with its lower end on the quadrant of the curve which is furthest removed from the focus. Shew that its length must be at least $\frac{3}{4}a + \frac{1}{4}a\sqrt{(1+8e^2)}$, where a is the semi-major axis and e the eccentricity.

47. A rigid bar without weight is suspended in a horizontal position by means of three equal, vertical, and slightly elastic rods to the lower ends of which are attached small rings A, B, and C through which the bar passes. A weight is then attached to the bar at any point G. Shew that, assuming that the extension or compression of an elastic rod is proportional to the force applied to stretch or compress it, and provided the rods remain vertical, the rod at B will be compressed, if G lie in the longer of the two arms AB, BC, and be at a distance from B greater than $(AB^2 + BC^2)/(AB \sim BC)$.

48. A number n of equal smooth spheres of weight W and radius r is placed within a hollow vertical cylinder of radius a, less than $2r$, open at both ends and resting on a horizontal plane. Prove that the least value of the weight W' of the cylinder in order that it may not be upset by the balls is given by

$$aW' = (n-1)(a-r)W \quad \text{or} \quad aW' = n(a-r)W,$$

according as n is odd or even.

49. Four equal smooth spherical balls of radius a are piled up within a hollow sphere which is the largest which can retain them in mutual contact, shew that its radius is $a(1 + 2\sqrt{1\overline{1}})$.

50. A set of equal frictionless cylinders, tied together by a fine string in a bundle whose cross section is an equilateral triangle, lies on a horizontal plane. Prove that if W be the total weight of the bundle and n the number of cylinders in a side of the triangle, the tension of the string cannot be less than $\dfrac{W}{4\sqrt{3}}\left(1+\dfrac{1}{n}\right)^{-1}$ or $\dfrac{W}{4\sqrt{3}}\left(1-\dfrac{1}{n}\right)$, according as n is an even or an odd number, and that these values will occur when there are no pressures between the cylinders in any horizontal row above the lowest.

51. A quadrilateral $ABCD$ has the sides DA, AB, BC equal and the angles DAB, ABC right angles, but AB and CD are not in the same plane. If forces acting along the four sides can be reduced to a couple, its axis will make with AB an angle

$$= \cos^{-1}\sqrt{\dfrac{CD^2 - AB^2}{CD^2 + 3AB^2}}.$$

52. Forces act along the edges BC, CA, AB, OA, OB, OC of a finite tetrahedron, represented in magnitude by λBC, μCA, νAB, $\lambda' OA$, $\mu' OB$, $\nu' OC$ respectively. Prove that they will be equivalent to a couple, if

$$\lambda' + \mu - \nu = \mu' + \nu - \lambda = \nu' + \lambda - \mu = 0.$$

53. Prove that the axis of the resultant of two given wrenches $(R_1 H_1)$ and $(R_2 H_2)$, the axes of which are inclined to each other at an angle θ, intersects the shortest distance ($2c$) between their axes at a point the distance of which from the middle point is

$$\dfrac{(R_1^2 - R_2^2)c + (H_1 R_2 - H_2 R_1)\sin\theta}{R_1^2 + R_2^2 + 2R_1 R_2 \cos\theta}.$$

CHAPTER III.

STATICS OF CONSTRAINED BODIES, ETC.

72. THE conditions of equilibrium which we have proved in the last Chapter apply to any rigid bodies whatsoever. If however the body considered be a *constrained* one, i.e. one that is not free to move in every way, as for instance one that can only turn about a fixed axis, we can obtain conditions of equilibrium which do not involve the forces of constraint.

73. Prop. *If a rigid body under the action of a system of coplanar forces, have one point in the plane of the forces fixed, it is a necessary and sufficient condition of equilibrium that the algebraical sum of the moments about the fixed point of the forces, excluding the force of constraint, be zero.*

For the force of constraint acts through the fixed point A, and therefore when there is equilibrium, the resultant of the remaining forces must act through A. But the algebraical sum of the moments of these remaining forces about any point is equal to the moment of their resultant, and therefore that about A vanishes. The condition is therefore a *necessary* one.

It is also *sufficient*. For if it hold, it can be shewn as in Art. 52 that A being a fixed point, the body is in equilibrium.

98   STATICS.

Ex. 1. A uniform rod which is 12 feet long and which weighs 17 lbs. can turn freely about a point in it, and the rod is in equilibrium when a weight of 7 lbs. is hung at one end. How far from that end is the point about which it can turn?   *Ans.* 4 ft. 3 in.

Ex. 2. $ABCD$ is a square: a force of 1 lb. acts from A to B, one of 4 lbs. from B to C, and one of 15 lbs. from D to C: if the centre of the square is fixed, find the force which, acting along DA, will maintain equilibrium.   *Ans.* 10 lbs.

Ex. 3. $ABCD$ is a square, of which the point A is fixed: a force of 2 lbs. acts along AB, one of 6 lbs. along AD, one of 10 lbs. along BD, and one of 3 lbs. along BC, find the force along DC which will maintain equilibrium.   *Ans.* $(5\sqrt{2}+3)$ lbs.

Ex. 4. A lever ABC, with a fulcrum B, one-third of its length from A, is divided into equal parts in D, E, and F. At C, D, and F, forces of 12 lbs., 8 lbs., and 6 lbs. respectively act vertically downwards, and at E a force of 16 lbs. acts vertically upwards. What force applied to A will cause equilibrium?   *Ans.* $21\frac{1}{2}$ lbs.

Ex. 5. A weightless lamina in the shape of a regular hexagon $ABCDEF$, is suspended from the middle point of AB: shew that it will be in equilibrium with the side AB horizontal, if weights of 3 lbs., 7 lbs., 3 lbs. and 5 lbs. are hung at C, D, E, and F respectively.

74. Prop. *If two points of a rigid body be fixed, so that it can only turn about the line joining them, it is a necessary and sufficient condition of equilibrium that the algebraical sum of the moments of the forces, excluding those of constraint, about the fixed line, be zero.*

If there is equilibrium, the algebraical sum of the moments of *all* the forces about any line is zero, and the moment of the force of constraint at each of the fixed points about the line joining them is zero: therefore the sum of the moments of the remaining forces, excluding those of constraint, about this line, is zero. It is therefore a *necessary* condition.

It can be proved as in Art. 54 that when the algebraical sum of the moments about any line is zero, there is equilibrium provided two points in the line be fixed. The condition is therefore *sufficient*.

75*. Prop. *If one point of a rigid body be fixed, the necessary and sufficient conditions of equilibrium are, that the algebraical sum of the moments of the forces about each of three lines through the fixed point, but not in the same plane, be zero.*

It can be shewn, as in the last proposition, that the conditions are *necessary*.

It can be shewn, as in Art. 55, that they are *sufficient*.

76. To obtain the forces of constraint at the fixed points in any of the cases considered in the last three propositions, we have only to apply the remaining conditions of equilibrium found in Chapter II.

77. As we shall often have to consider the case of bodies, such as rods, which are connected by means of *hinges* or *joints*, it will be well to consider what a *hinge* is. We shall consider *smooth* hinges only.

The connection may be supposed to be made in several ways. A point of one body may be connected with one of the other body by a very short *string*. Or one body may end in a very small *ball* or *pivot*, which works in a corresponding small *socket* or *ring* in the other body, so that there is contact at only one point. Or we may suppose each body to end in a small ball, which works in a corresponding socket of a small separate body. In each of these cases there is no restriction on either body, except that the two ends must be in contact; the action on each at the common point must pass through this point, but will adapt itself in magnitude and direction so as to maintain equilibrium, if possible.

If three or more bodies are connected by one joint, we may suppose the connection to be made by each having a very short string attached to it, and the strings to be knotted together. Or we may suppose each to end in a small smooth ball, which works in a corresponding socket in a small separate body.

78. In the construction of materials it is often desirable to ascertain the internal forces between one portion of a body and the adjacent portion. When all these are known, we are able to adapt the strength of each part to the force it has to sustain. For instance, if we know that the tension at one point of a chain is

always half that at another, the thickness of the chain at the former point need only be half that at the latter; a saving in material and in weight is thus effected.

We have learnt that when a body is in equilibrium, the forces exerted on any portion of it by the adjacent portions counteract the remaining forces acting on the portion in question. As, however, there is an infinite number of systems of forces, each of which counteracts a given system, we cannot as a rule determine which system is the one actually exerted, without going beyond the limits of Elementary Statics. If, for instance, a rope composed of several fibres be taut, though we may know the tension of the rope itself, i.e. the sum of the tensions of the different fibres, we cannot say how it is distributed among them. This can only be ascertained when the elasticity of each fibre is known.

When a beam is merely *stretched*, i.e. when the external forces all act *along* it, the only internal forces called into play will be between particles arranged in lines along the beam. If then the beam be supposed to consist of two parts A and B, the action of B on A will be the sum of the forces exerted by particles of B on the adjacent particles of A, all such forces being in the same direction along the beam. This action is equal to the resultant of the forces acting on the portion A, and which are also external to the beam. It is clear that the greater this action becomes, the more likely is the beam to be pulled asunder at the point of junction of A and B; the action therefore measures the *tendency* of the beam *to break* at that point.

79. When the external forces on the beam are not all along it, the action of one portion on another is not so simple as in the above case. Take the following case. Let $ABCD$ be a rectangular beam which is firmly fixed at the end AB in a vice; along DC let a force S be applied: it will of course be perpendicular to the beam. Consider

the equilibrium of the portion $CDPQ$, where PQ is an imaginary section perpendicular to the beam's length.

The forces in action are S and the innumerable forces due to $ABQP$, acting at every point of the section PQ.

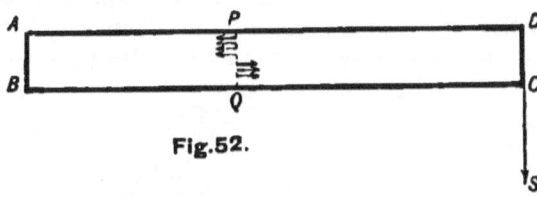

Fig. 52.

Let the latter be resolved along PQ and at right angles to it: the sum of the former components must be equal and opposite to S, and will with it form a couple. The components perpendicular to PQ must therefore be equivalent to a couple, equal and opposite in sign to the former. This shews that the forces near P must be in the direction PA, and those near Q in the opposite direction: and therefore that the tendency of S is to stretch the fibres near P and crush those near Q. It must follow too, that the magnitudes of the components perpendicular to QP depend on the moment of S about Q, and not on the magnitude of S simply. Hence the greater the moment of S about Q the more likely are the fibres along PQ to give way and the rod to bend at PQ.

Since PQ is supposed small compared with QC, the numerical sum of the forces along PQ must be very much greater than S, i.e. a force is far more likely to bend a rod, when applied at right angles to it, than to pull it asunder when applied along it.

Similar reasoning will apply to a beam under the action of any system of forces. We can shew that the *tendency to bend* at any point is measured by the algebraical sum of the moments about that point, of the forces external to the rod and acting on one of the parts into which the beam is divided by the point. This *tendency to bend* is also termed the *bending moment*.

Ex. 1. A light beam is supported in a horizontal position at its ends, and a weight w is hung from its middle point. Find the bending moment at a point distant x from one end. *Ans.* $\tfrac{1}{2}wx$.

Ex. 2. If a heavy uniform rod be supported at its middle point, shew that the bending moment at any point varies as the square of its distance from the nearer end.

Ex. 3. A uniform rod AB of weight w and length a is supported in a horizontal position at A and B; from a point distant x from A a weight w' hangs: find the bending moment at a point distant y from A.

Ans. $w \cdot \dfrac{y(a-y)}{2a} + w' \cdot \dfrac{(a-x)y}{a}$, if y is $< x$,

$w \cdot \dfrac{y(a-y)}{2a} + w' \cdot \dfrac{(a-y)x}{a}$, if y is $> x$.

Ex. 4. A uniform rod of weight w and length a, can turn freely about a hinge at one end, and rests with its other end against a smooth vertical wall, distant b from the hinge. Prove that the bending moment at a point whose distances from the two ends are x, y, respectively, is $\tfrac{1}{2}wxyba^{-2}$.

Ex. 5. A uniform rod of length a rests horizontally on two pegs, one at one end of the rod, find where the other peg must be placed so that the bending moment at a point distant x from the first peg may be zero.

Ans. $a^2/(2a-x)$ from the first peg.

80. When each of the bodies forming a system in equilibrium is acted on by forces that reduce to three, the problem of finding the position of each of the bodies

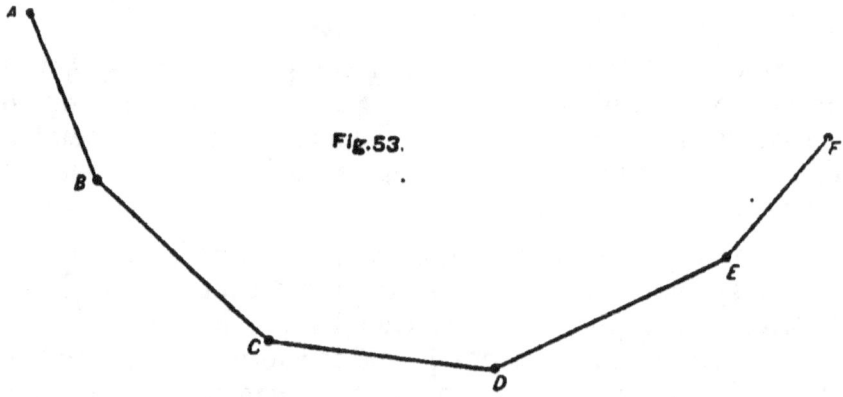

Fig. 53.

or of ascertaining the different forces, can often be easily solved by constructing a series of triangles each of which is the *triangle of forces* corresponding to one of the bodies. For instance, let us consider the case of a number of particles of equal weight fastened at intervals along a weightless string, the ends of which are attached to fixed points. Let A, B, C, D &c. be the positions of the particles, when in equilibrium. Any particle, B for instance, is kept in equilibrium by three forces, its weight vertically downwards, and the tensions of the strings BA, BC. Draw a triangle Obc, having its sides ba, aO, Ob respectively parallel to the lines of action of these forces: then by the

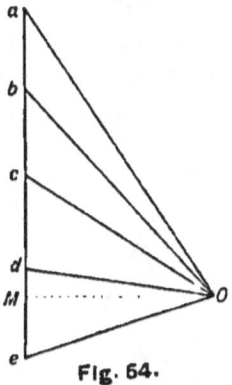

Fig. 54.

triangle of forces these lines are proportional to the forces, to whose directions they are parallel: i.e. weight of A : tension of AB : tension of $BC = ab : aO : Ob$. Produce ab downwards, and mark off bc, cd, de &c., each equal to ab; join Oc, Od, Oe &c. Then Ob, bc represent in every way the tension of BC, on C, and the weight of C respectively, so that cO must represent the tension of CD. Similarly dO represents the tension of DE, eO that of EF, and so on.

Draw OM perpendicular to abc: then the tangents of the angles that aO, bO, cO &c. make with the horizon are

$$\frac{aM}{OM}, \quad \frac{bM}{OM}, \quad \frac{cM}{OM}, \quad -\frac{Me}{OM};$$

hence the tangents of the angles which the strings make with the horizon form an arithmetic series. Also the horizontal resolved part of the tension of each string is represented by OM, and is therefore the same for all.

Such a figure which is drawn to enable us to solve the problem is called a *Force-Diagram*.

The above results can be obtained very easily by equating to zero the algebraical sums of the resolved parts in a horizontal and vertical direction of the forces that act on each particle separately.

81. This '*Graphic*' method can be applied to prove the following important proposition.

Prop. *If a weightless string be stretched across a smooth surface, the tension is everywhere the same.*

Let $ABCD$ &c. be the string: then any small portion

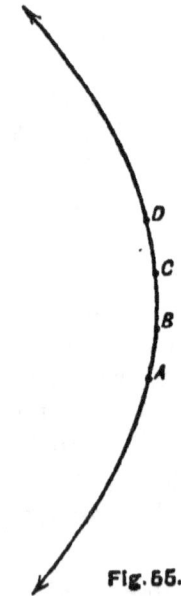

Fig. 65.

of it AB is kept in equilibrium by the tensions at its ends, and the resultant of the pressures of the surface on it: as the pressures along AB all act along the normals to the surface at the corresponding points, their resultant's direction must lie somewhere between the normals at A and B.

Draw Oa, Ob, Oc, Od, Oe &c. parallel to the tensions at

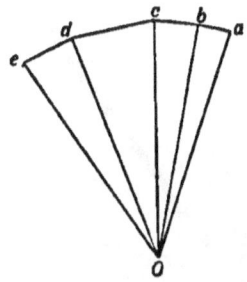

Fig. 56.

A, B, C, D &c. respectively: and also ab, bc, cd, &c. parallel to the resultants of the pressures on AB, BC, CD &c. Then by the triangle of forces, each line represents the magnitude of the force to whose direction it is parallel. Since the resultant pressure on AB has a direction between the normals at A and B, and these ultimately, when AB is taken indefinitely small, make indefinitely small angles with one another, ab makes with the normals at A and B very small angles, i.e. makes with Oa, Ob, which are parallel to the tangents at A, B, angles ultimately equal to right angles. Hence the difference between Oa, Ob must be of the second order of small quantities, similarly those between Ob and Oc, Oc and Od &c. are of the second order, i.e. Oa, Ob, Oc &c. and the tensions they represent are all equal.

ILLUSTRATIVE EXAMPLES.

Ex. 1. OA, AB are two uniform beams loosely jointed at A, the former being moveable about a hinge at O. A string attached to B passes over a fixed smooth pully and supports a weight P. If in the position of equilibrium the beams are equally inclined to the vertical, the string will make an angle $\cos^{-1}\left(\dfrac{W+3W''}{4P}\right)$ with the vertical, where W, W'' are the weights of the beams.

Let α be the inclination of either of the beams to the vertical, and θ that of the string.

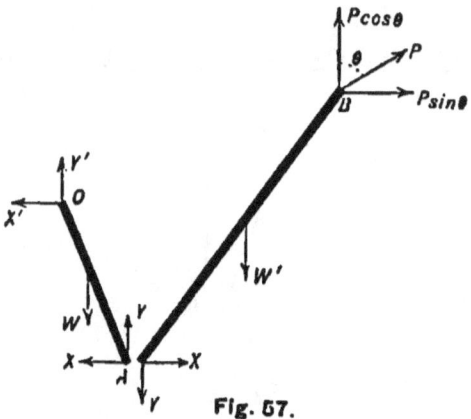

Fig. 57.

Resolve the tension of the string (P) at B into two forces $P \cos \theta$ vertically, and $P \sin \theta$ horizontally.

Let $2a$, $2b$ be the respective lengths of OA, AB.

From the equilibrium of both rods together by taking moments about O, we have

$W a \sin \alpha + W''(2a \sin \alpha + b \sin \alpha) - P \cos \theta (2a \sin \alpha + 2b \sin \alpha)$

$\qquad + P \sin \theta (2b \cos \alpha - 2a \cos \alpha) = 0 \ldots (1).$

Taking moments about A for the equilibrium of AB

$W''b \sin \alpha + P \sin \theta \cdot 2b \cos \alpha - P \cos \theta \cdot 2b \sin \alpha = 0 \ldots\ldots\ldots\ldots(2).$

Subtracting

$\qquad W a \sin \alpha + 2W''a \sin \alpha - 2Pa \sin(\theta + \alpha) = 0,$

or $\qquad (W + 2W') \sin \alpha = 2P \sin(\theta + \alpha) \ldots\ldots\ldots\ldots\ldots(3),$

from (2) $\qquad W'' \sin \alpha = 2P \sin(\alpha - \theta) \ldots\ldots\ldots\ldots\ldots\ldots(4).$

Adding equations (3) and (4) we have

$\qquad (W + 3W'') \sin \alpha = 4P \sin \alpha \cos \theta,$

$\qquad \therefore \cos \theta = (W + 3W')/4P$

If the stresses at O and A be resolved horizontally and vertically, as shewn in the figure, we can determine them as follows:

Resolving horizontally and vertically for the equilibrium of OA

$$X' + X = 0 \quad \ldots\ldots(5),$$
$$Y' + Y - W = 0 \quad \ldots\ldots(6).$$

Resolving horizontally and vertically for AB,

$$X + P \sin \theta = 0 \quad \ldots\ldots(7).$$
$$Y + W - P \cos \theta = 0 \quad \ldots\ldots(8).$$

Equations (5), (6), (7) and (8) completely determine X, Y, X' and Y'.

Ex. 2. An equilateral pentagon consisting of five freely-jointed rods is hung up with one side horizontal; shew that the inclination (θ) of either of the upper rods to the vertical is given by the equation

$$\sin \theta + 6 \sin^2 \theta + 8 \sin^3 \theta - 8 \sin^4 \theta = \tfrac{1}{4}.$$

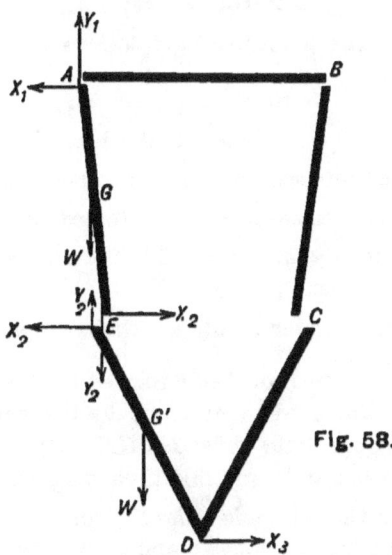

Fig. 58.

Let AB be the fixed rod. Let ϕ be the inclination of CD and DE to the vertical.

(The rods are drawn separate to make the figure clearer.)

Let W be the weight of each rod, $2a$ its length.

Let the stresses on the different rods at the joints be resolved horizontally and vertically: the magnitude of these stresses are indicated in the figure. The stress at D is entirely horizontal, as the rods CD, DE are symmetrical with respect to the vertical.

From the equilibrium of AE, resolving horizontally we have
$$X_1 = X_2 \quad \quad \quad \quad \quad \quad \quad \quad (1).$$
Taking moments about A
$$Wa \sin \theta + Y_2 \cdot 2a \sin \theta - X_2 \cdot 2a \cos \theta = 0 \quad \quad (2).$$
From the equilibrium of ED, resolving horizontally
$$X_2 = X_3 \quad \quad \quad \quad \quad \quad \quad \quad (3).$$
Resolving vertically
$$Y_2 = W \quad \quad \quad \quad \quad \quad \quad \quad (4).$$
Taking moments about E
$$Wa \sin \phi - X_3 \cdot 2a \cos \phi = 0 \quad \quad \quad \quad (5).$$
Substituting from (1), (3) and (4) in (2) and (5),
$$2X_1 \cos \theta = 3W \sin \theta \quad \quad \quad \quad \quad (6),$$
$$W \sin \phi = 2X_1 \cos \phi \quad \quad \quad \quad \quad (7),$$
$$\therefore \cot \theta = 3 \cot \phi \quad \quad \quad \quad \quad (8).$$

Since the sum of the horizontal projections of AE, ED, DC, CB is equal to AB,
$$4a \sin \theta + 4a \sin \phi = 2a,$$
$$\therefore \sin \theta + \sin \phi = \tfrac{1}{2} \quad \quad \quad \quad \quad (9).$$

By eliminating (ϕ) between (8) and (9) we obtained the required result.

By substituting the value of θ just obtained in (6), we determine X_1 and by resolving vertically for the equilibrium of AE, we obtain another equation which determines Y_1.

The stresses at the angular points are thus completely determined.

Ex. 3. Six equal heavy rods freely jointed at the ends form a regular hexagon $ABCDEF$, which when hung up by the point A is kept from altering its shape by two light rods BF, CE. Prove that the thrusts of the rods BF, CE are as 5 to 1, and find their magnitudes.

We shall suppose that there is a light pivot at B, to which the three rods AB, BF, and BC are attached; and that a similar arrangement is made at C.

Let W be the weight of each rod, $2a$ its length.

Since the rods BF, CE are only acted on by the stresses at their ends, these stresses must be along the rods, i.e. horizontal, let them be S and T respectively.

Since the rod BC is acted on by its weight along its length and the stresses at B and C, these latter forces must also act along BC (Art. 61), i.e. vertically.

STATICS OF CONSTRAINED BODIES, ETC. 109

From symmetry the stress at D on CD is horizontal.

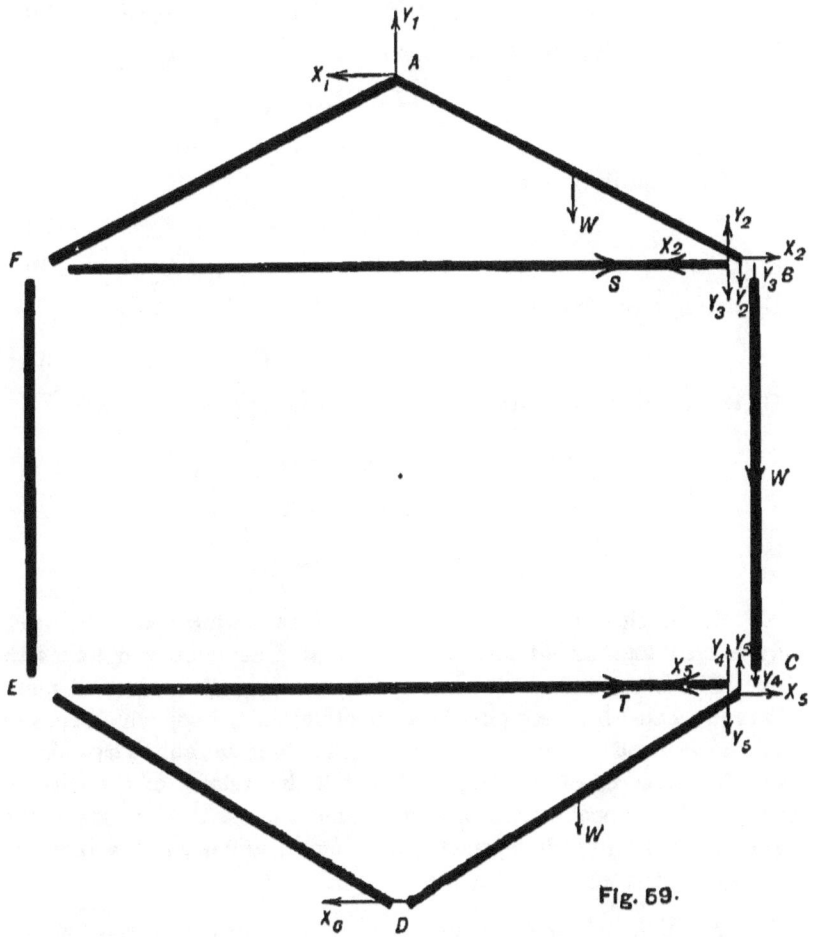

Fig. 59.

Let the stresses resolved horizontally and vertically at A, B, C and D be those shewn by the figure.

From the equilibrium of AB, by taking moments about A,

$$W \cdot a \cdot \tfrac{1}{2}\sqrt{3} + Y_2 \cdot 2a \cdot \tfrac{1}{2}\sqrt{3} - X_2 \cdot 2a \cdot \tfrac{1}{2} = 0 \dots\dots\dots\dots(1).$$

From the equilibrium of the pivot B, resolving horizontally and vertically,

$$S = X_2 \dots\dots\dots\dots\dots\dots\dots\dots(2)$$

and

$$Y_2 = Y_3 \dots\dots\dots\dots\dots\dots\dots\dots(3).$$

From the equilibrium of BC,

$$Y_3 - W - Y_4 = 0 \quad \dots \dots (4).$$

From the equilibrium of the pivot C,

$$T = X_5 \quad \dots \dots (5),$$

$$Y_4 = Y_5 \quad \dots \dots (6).$$

From the equilibrium of CD,

$$X_5 = X_6 \quad \dots \dots (7),$$

$$Y_5 = W \quad \dots \dots (8).$$

By taking moments about C,

$$W \cdot a \cdot \tfrac{1}{2}\sqrt{3} - X_6 \cdot 2a \cdot \tfrac{1}{2} = 0 \quad \dots \dots (9).$$

By substituting from the other equations in (1) and (9), we have

$$W \cdot \tfrac{1}{2}\sqrt{3} + 2W \cdot \sqrt{3} - S = 0,$$

$$W \cdot \sqrt{3} - 2T = 0,$$

$$\therefore S = 5W \cdot \tfrac{1}{2}\sqrt{3} = 5T.$$

Ex. 4. A gipsy's tripod consists of three uniform straight sticks freely hinged together at one end. From this common end hangs the kettle. The other ends of the sticks rest on a smooth horizontal plane, and are prevented from slipping by a smooth circular hoop which encloses them and is fixed to the plane. Shew that there cannot be equilibrium unless the sticks be of equal length; and if the weights of the sticks be given (equal or unequal) the bending moment of each will be greatest at its middle point, will be independent of its length, and will not be increased on increasing the weight of the kettle.

Let OA, OB, OC be the three rods, P, Q, R their respective weights acting at their middle points. Let X, Y, Z be the vertical stresses at A, B and C, and X', Y', Z' the horizontal stresses.

Draw OH vertically downwards.

The three forces acting on OA, viz. P and the resultant stresses at O and A, must be in one plane (Art. 61) the vertical plane containing OA, i.e. OAH.

X' the horizontal stress at A must therefore act along AH; similarly Y' and Z' act along BH and CH respectively.

STATICS OF CONSTRAINED BODIES, ETC.   111

But these horizontal stresses act along the normals to the circle ABC, so that H must be the centre of that circle. The lines HA, HB, HC

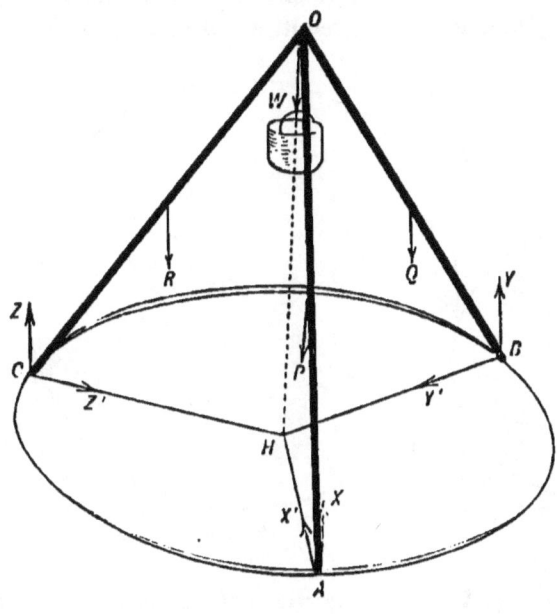

Fig. 60.

must therefore be equal to one another, and also OA, OB, OC to one another.

Let $2l$ be the length of each rod, θ its inclination to the horizon.

Taking moments about O for the equilibrium of OA, we have

$$X \cdot 2l\cos\theta - Pl\cos\theta - X' \cdot 2l\sin\theta = 0$$

$$\therefore 2X - P - 2X'\tan\theta = 0.$$

The bending moment at a point on OA distant x from A

$$= Xx\cos\theta - X'x\sin\theta - \frac{Px}{2l} \cdot \frac{x}{2}\cos\theta = \frac{P\cos\theta}{4l}(2lx - x^2),$$

$$= \frac{P\cos\theta}{4l}\{l^2 - (l-x)^2\}.$$

This is clearly a maximum, when $x = l$, i.e. the bending moment is greatest at the middle point, where it is equal to $\frac{1}{4}Pl\cos\theta$, or $\frac{1}{4}Pr$, where r is the radius of the hoop, i.e. is independent of l and W.

112   STATICS.

Ex. 5. An elastic band binds together any number of smooth right cylinders so that each cylinder touches only two others. Prove that if lines be drawn from a point parallel and proportional to the pressures between the cylinders, their extremities will lie on a circle.

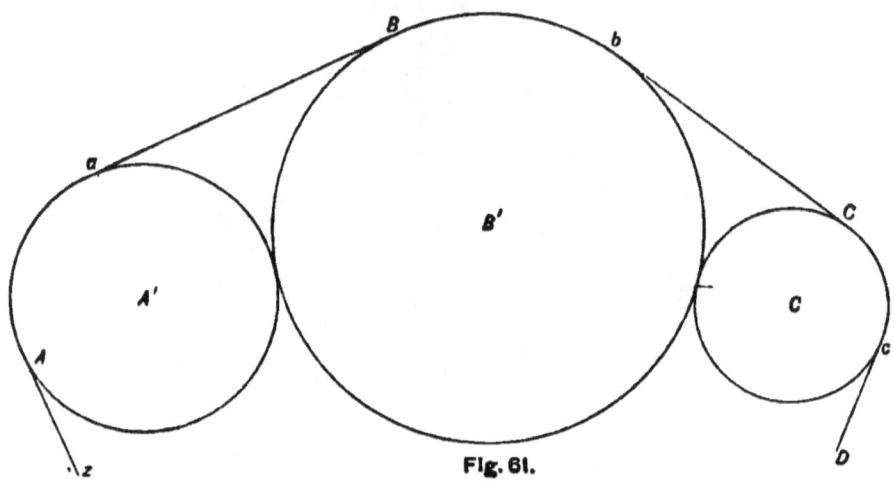

Fig. 61.

Let Aa, Bb, Cc, &c. be the portions of the band in contact with the cylinders A', B', C', &c.

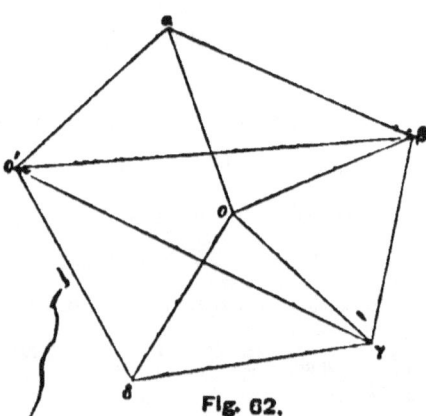

Fig. 62.

From any point O draw a number of equal straight lines Oa, $O\beta$, $O\gamma$, $O\delta$, &c. respectively parallel to the portions of the band zA, aB, bC, cD, &c. These lines will therefore represent the tensions along the corresponding portions of the band.

Join $\alpha\beta$, $\beta\gamma$, $\gamma\delta$, &c.

By the triangle of forces, $\alpha\beta$ represents the resultant action of Aa on the cylinder A'. Similarly $\beta\gamma$, $\gamma\delta$, &c. represent the resultant actions of the band on the cylinders B', C', &c. respectively.

Through β draw $\beta O'$ parallel to the normal common to A' and B': through γ draw $\gamma O'$ parallel to the normal common to B' and C'. Join $\delta O'$, $\epsilon O'$, &c. By the triangle of forces $O'\beta$ and $\gamma O'$ represent the pressures of the cylinders A' and C' on B'.

Therefore $O'\gamma$, $\gamma\delta$ represent two of the forces on the cylinder C', so that $\delta O'$ must represent the third, which is the pressure due to D'. Similarly it can be shewn that $O'\alpha$, &c. represent the pressures between the other pairs of cylinders.

Hence from O' straight lines $O'\alpha$, $O'\beta$, $O'\gamma$, &c. have been drawn representing in magnitude and direction the pressures between the cylinders, and their extremities α, β, γ lie on a circle whose centre is O, since $O\alpha$, $O\beta$, $O\gamma$, &c. are all equal.

(The cylinders are not necessarily circular.)

EXAMPLES.

1. Two uniform heavy rods, each of length a and jointed together by a smooth hinge, are placed symmetrically over two pegs at a given distance b apart in a horizontal line; prove that in the position of equilibrium each rod is inclined to the horizon at an angle $\cos^{-1}(b/a)^{\frac{1}{3}}$.

2. Three equal uniform rods, AB, BC, CD, of the same material and thickness, are jointed at B and C. If they are supported in a horizontal plane by smooth pegs placed under AB and CD, shew that the distance between either peg and the nearest joint is one-third the length of a rod.

3. A uniform heavy rod of length $2b$ and weight W can turn freely about one end. To this end is attached a string of length $l(<2b)$, which supports a sphere of radius a and weight W'. When the system is in equilibrium with the rod resting against the sphere, the rod makes an angle θ with the horizontal; shew that $\tan\theta - \tan\alpha = Wb/W'a$, where $l = a(\sec\alpha - 1)$, and $l^2 + 2al$ is $<4b^2$.

4. A uniform heavy rod hangs by light inextensible strings, attached to its ends, and also to the ends of another uniform rod, which can turn about a pivot at its middle point. Prove that, when there is equilibrium, either the rods or the strings are parallel.

5. Prove that the angular points of a funicular polygon, in which the weights are equal and also the horizontal distances between them, lie on a parabola.

6. Two rods AC, BC, of equal uniform thickness are jointed at C, and the ends A and B are fixed at two points in the same vertical line. Prove that the direction of the action at the joint C bisects the angle ACB: and if $AB^2 = 4AC \cdot BC$, shew that its magnitude is equal to a quarter of the difference of the weights of the rods.

7. A chain formed of rods of equal weight jointed together is hung up by its two ends and rests under the action of gravity. Shew that, if lines be drawn from a point representing the actions at the hinges, their ends lie on a straight line.

8. A rhombus is formed of four similar uniform rods connected by smooth hinges at their extremities, and two of these rods rest upon two smooth pegs in the same horizontal line: determine the position in which the rhombus will rest with one of its diagonals vertical.

9. Two uniform rods AB, AC, of lengths a, b respectively, are of the same material and thickness and smoothly jointed at A. A rigid weightless rod of length l is jointed at B to AB and its other end D is fastened to a smooth ring sliding on AC. The system is hung over a smooth peg at A: shew that AC makes with the vertical an angle

$$\tan^{-1} \frac{al}{b^2 + a\sqrt{(a^2 - l^2)}}.$$

10. A regular tetrahedron consists of six rigid bars without weight. It is suspended from one angular point, and from the other three equal weights W are hung: find the strain on each of the horizontal edges.

11. A beam AB of length a and weight w rests horizontally on two smooth pegs, whose distances from A and B respectively are $\frac{1}{3}a$ and $\frac{1}{4}a$: if from A a weight $5w$ is hung, and from B, $\frac{7}{2}w$, shew that the bending moment is greatest at the peg next A, and find its magnitude.

12. Two heavy uniform rods AB, BC, weights P and Q, are connected by a smooth joint at B. The ends A and C slide by means of small smooth rings on two fixed rods each inclined at an angle α to the horizon. If θ and ϕ be inclinations of the rods AB, BC respectively to the horizon, shew that

$$(P+Q)\tan\alpha = Q\cot\theta = P\cot\phi.$$

13. At what distance from the foot of an upright post must a rope of given length be attached, in order that a given force applied to the other end may produce the greatest bending moment at the foot of the post?

14. Four uniform rods AB, BC, CD, DA freely jointed at their ends so as to form a quadrilateral rest on a smooth horizontal table. They are connected together by an endless elastic string passing through small smooth rings at their middle points. Prove that in the position of equilibrium, the harmonic means of the segments into which each diagonal is divided by the other are equal.

15. A heavy uniform rod of weight W and length $2a$ can turn freely about a hinge at one end; a ring of weight w, which slides along the rod, is connected with a point in the same horizontal plane as the hinge, by means of a string whose length c is equal to the distance between the point and the hinge. Show that in the position of equilibrium the angle θ which the rod makes with the horizon is given by the equation
$$2wc \cos 2\theta + Wa \cos \theta = 0.$$

16. Two equal uniform ladders of length l, freely jointed at A, are connected by a rope PQ and rest equally inclined to it on a smooth horizontal plane; a man of weight W goes a distance b up one of the ladders: prove that the tension of the rope is $\dfrac{Wb+wl}{2a} \cdot \dfrac{c}{\sqrt{(a^2-c^2)}}$, if $w=$ weight of each ladder, $2c=$ length of the rope, and $AP=AQ=a$.

17. AB, BC, CD are three equal rods freely jointed at B and C. The rods AB, CD rest on two pegs in the same horizontal line so that BC is horizontal. If α be the inclination of AB, and β that of the reaction at B to the horizon, prove that $3\tan\alpha \tan\beta = 1$.

18. Two equal rods can move in a vertical plane about an axis through their middle points. The lower ends of the rods are connected by a weightless elastic string, and a circle of weight W rests between the rods above the joint. The radius of the circle and the unstretched length of the string are each equal to half the length of either rod, and the rods are at right angles when the system is in equilibrium; prove that Young's modulus for the string is $(\sqrt{2}-1)$.

19. Four equal uniform rods, each of weight W, are jointed so as to form a rhombus $ABCD$: the system rests on a horizontal plane, with AC vertical, B, D are connected by a light string: shew that its tension is $2W \tan \tfrac{1}{2} BAD$, and find the actions on the rods at A and C.

20. A triangular lamina ABC is moveable in its own plane about a point in itself: forces act on it along and proportional to BC, CA, BA. Prove that if these do not move the lamina the point must lie in the straight line which bisects BC and CA.

21. Five rods are jointed so as to form a regular pentagon $ABCDE$ and are suspended from A. Two strings connect C with the middle point of AE, and D with the middle point of AB. Determine their tensions.

22. Seven equal and similar uniform rods AB, BC, CD, DE, EF, FG, GA are freely jointed at their extremities and rest in a vertical plane supported by rings at A and C, which are capable of sliding on a smooth horizontal rod: prove that, θ, ϕ, ψ being the angles which BA, AG, GF make with the vertical, $\tan\theta = 4\tan\phi = 2\tan\psi$.

23. Four rods jointed at their extremities form a quadrilateral, which may be inscribed in a circle: if they be kept in equilibrium by two strings joining the opposite angular points, show that the tension of each string is inversely proportional to its length, the weights of the rods being neglected.

24. A series of particles are knotted on an endless string, forming a closed polygon, and are in equilibrium under the action of given forces applied to the particles. Shew that the tensions of the string may be represented in direction and magnitude by means of straight lines drawn from a point to the angular points of the polygon of forces.

25. Three uniform rods AB, BC, CD, lengths $2c$, $2b$, $2c$, rest symmetrically on a smooth parabolic arc, lat. rec. $= 4a$, whose axis is vertical and vertex upwards. There are hinges at B and C, and all the rods touch the parabola. If W be the weight of either slant rod, shew that its pressure against the parabola is $W a^2 c / \{b(a^2 + b^2)\}$.

26. Four equal uniform rods are freely jointed at their extremities so as to form a square, and the middle point of one side is joined by three strings to the middle points of the other three sides.

(1) If the square be laid on a smooth table, prove that the tensions of two of the strings will be equal: and, given the magnitudes of the three tensions, find the actions at the joints.

(2) If the square be hung up by one corner, prove that the difference between two of the tensions will be four times the weight of a rod.

27. Two equal rods AB, BC, of length $2a$, are connected by a free hinge at B: the ends A and C are connected by an inextensible string of length l: the system is suspended from A: prove that, in order that the angle AB makes with the vertical may be the greatest possible, l must be equal to $\tfrac{4}{3}a\sqrt{3}$.

28. Six equal and uniform heavy rods are hinged together so as to form a hexagon: it is placed with one side on a horizontal plane and is kept in the shape of a regular hexagon by means of a string fastened to the middle points of the two sides adjacent to the base: find the tension of the string and the stresses at the hinges.

29. A parallelogram formed of four rods of uniform material and thickness, jointed at their ends, is suspended from one point, which is connected with the opposite point by a string of such a length that the figure is rectangular: prove that the tension of the string is half the weight of the four rods, and that the direction of the stress between the rods at either of the joints not connected by the string bisects the angle between them.

30. A heavy uniform rod of length $2a$ turns freely on a pivot at a point in it, and suspended by a string of length l fastened to the ends of the rod hangs a bead of equal weight which slides on the string. Prove that the rod cannot rest in an inclined position unless the distance of the pivot from the middle point of the rod be less than a^2/l.

31. A number of equal weightless rods are freely jointed and assume the form of a regular polygon when subjected to a system of stresses at each joint, all emanating from a point on the circumscribing circle. Shew that, if from a point radii be drawn to represent in magnitude and direction the stresses in the rods, and a polygon be constructed so that its sides taken in order represent the system of applied stresses, then the polygon will be equiangular and described about a parabola, and further the angular points of the polygon will all lie on a hyperbola.

32. Two equal beams AB, AC, connected by a hinge at A, are placed in a vertical plane with their extremities B, C resting on a horizontal plane; they are kept from falling by strings connecting B and C with the middle points of the opposite sides; shew that the ratio of the tension of each string to the weight of each beam is $\tfrac{1}{8}\sqrt{(8\cot^2\theta + \cosec^2\theta)}$, where θ is the inclination of either beam to the horizon.

33. A trapezium $ABCD$ is formed of four rods joined by hinges at their extremities: BC, AD are equal, and the framework is suspended by a string attached to the middle point of AB. Determine completely the stresses at A and D.

If $AB = AD = BC = \frac{1}{2}CD$, prove that

the stress at A : the stress at $D = \sqrt{19} : \sqrt{7}$.

34. A number of light rigid rods are loosely jointed together at their extremities so as to form a closed polygon, and a force applied to each side perpendicular and proportional to it, their lines of action meeting in a point; prove that, if equilibrium be maintained, the polygon will be inscribable in a circle, and if S be the point through which the forces act, O the centre of the circumscribed circle, and SO be produced to S' so that SS' is bisected in O, the stress at any angular point of the polygon will be perpendicular and proportional to the distance of the point from S'.

35. n equal uniform rods, each of weight W and length l, are jointed so as to form symmetrical generators of a cone whose semi-vertical angle is a, the joint being at the vertex of the cone. The rods are placed with their other ends in contact with the interior of a sphere whose radius is r, so that the axis of the cone is vertical, and a weight W' is hung on it at the joint. Shew that

$$\cos a = \frac{\sqrt{(r^2 - l^2)}.(nW + 2W')}{l\sqrt{(3n^2W^2 + 4nW'W)}},$$

and find the action at the joint on each rod.

36. A fire-screen holder with any number of unequal weightless arms projects horizontally from a chimney-piece. Shew that, if the ends of the arms all lie on a circle, the axes of the couples at the hinges all pass through one point.

37. A regular octahedron is formed of 12 uniform rods jointed together at the ends. Along the three diagonals are stretched strings whose tensions are T_1, T_2, T_3. Shew that the thrusts along the rods, joining the ends of the diagonals the tensions along which are T_1, T_2, are

$$\tfrac{1}{4}\sqrt{2}\,(T_1 + T_2 - T_3).$$

Prove also that, if the four diagonals of a cube be treated in a similar way, equilibrium is not possible unless the tensions are all equal.

38. Three uniform heavy rods of the same material (lengths $2a$, $2b$, $2c$, respectively) hinged together at B and C rest on a vertical circle of radius r, the whole system being in one vertical plane, and such that BC is horizontal. Find the stresses at the hinges, and prove that

$$(a^2 \cos^2 \theta - c^2 \cos^2 \phi + b^2 + 2bc) \cot \tfrac{1}{2} \theta$$
$$= (c^2 \cos^2 \phi - a^2 \cos^2 \theta + b^2 + 2ab) \cot \tfrac{1}{2} \phi$$
$$= (a+b+c)r,$$

where θ and ϕ are the acute angles made with the horizon by AB and CD respectively.

39. Three equal heavy rods, in the position of the three edges of an inverted triangular pyramid, are in equilibrium with their lower ends attached to a joint about which each rod can turn freely, and their upper ends connected by strings each of length equal to half that of a rod. Prove that the tension of a string is to the weight of a rod as $1 : \sqrt{11}$.

40. A rhombus is formed of four rods of length a, hinged together. Two opposite rods are supported in a vertical plane by two smooth pegs which are separated by a horizontal distance h and vertical distance k. Shew that the product of the horizontal distances of either peg from the ends of the nearer unsupported rod is $\tfrac{1}{4}(k^2 - 2ah + h^2)$, and that there is no bending moment round a point in either supported rod, whose distance from its supporting peg is three times the shorter of the distances of that peg from an unsupported rod.

41. Four equal uniform rods are jointed freely together so as to form a rhombus: this is suspended by one of the angular points, and a sphere of weight equal to twice that of the rhombus is balanced inside it so as to prevent it from collapsing; shew that, if the radius of the sphere be to the length of a rod in the ratio $5 : 8\sqrt{3}$, the rods will, in equilibrium, make each an angle $\tfrac{1}{6}\pi$ with the vertical.

42. $ABCD$ is a quadrilateral formed by four uniform rods of equal weight loosely jointed together. If the system be in equilibrium in a vertical plane with the rod AB supported in a horizontal position, prove that $2 \tan \theta = \tan a \sim \tan \beta$, where a, β are the angles at A and B, and θ is the inclination of CD to the horizon: also find the stresses at C and D, and prove that their directions are inclined to the horizon at the angles $\tan^{-1} \tfrac{1}{2}(\tan \beta - \tan \theta)$ and $\tan^{-1} \tfrac{1}{2}(\tan a + \tan \theta)$ respectively.

43. Four equal rods are joined together so as to form a rhombus $ABCD$, lying upon a smooth horizontal plane, and elastic strings AC, BD of the same substance are stretched along the diagonals: if a be the length of a side of the rhombus, and if the natural lengths of the strings be $\frac{1}{2}a$ and $\frac{1}{2}a(\sqrt{2}-1)$, find the angles of the rhombus when there is equilibrium.

44. Seven rods are freely jointed together to form a regular heptagon $ABCDEFG$. A weightless rigid rod connects BG, and two equal strings connect G with D and B with E, and the whole system is suspended by the point A. Find the tension of the strings.

45. A frame $ABCD$ is formed of four rods each of length a freely jointed together: it rests with AC vertical and the rods BC, CD in contact with fixed frictionless supports E, F, in the same horizontal line at a distance c apart, the joints B, D being kept apart by a rod of length b. Shew that, when a weight W is placed on the highest point A, it produces in BD a thrust of magnitude
$$\frac{W(2a^2c - b^3)}{b^2(4a^2 - b^2)^{\frac{1}{2}}}.$$
Examine the case when $b = (2a^2c)^{\frac{1}{3}}$.

46. A door is moveable about its line of hinges which is inclined at an angle a to the vertical; shew that the couple necessary to keep it in a position inclined at an angle β to its position of equilibrium is proportional to $\sin a \sin \beta$.

47. Three equal heavy rods AB, BC, CD are jointed to each other at B and C and to fixed points at A and D, where AD is horizontal and equal to the length of a rod. Shew that the horizontal couple required to turn the rod BC through an angle θ is $BC . W \sin \frac{1}{2}\theta$, where W is the weight of each rod.

48. The lid $ABCD$ of a cubical box, moveable about hinges at A and B, is held at a given angle to the horizon by a horizontal string connecting C with a point vertically over A: find the pressure on each hinge.

CHAPTER IV.

CENTRES OF MASS.

82. We have seen (Art. 59) that the resultant of *two* parallel forces, P, Q, acting at fixed points A, B respectively, is equal to their algebraical sum, and acts along a line parallel to the line of action of either: also that its line of action cuts AB at a fixed point, whose position depends solely on the relative magnitude of P and Q and not on their direction. So too, if we have *a number of* parallel forces acting at fixed points, their resultant is equal to their algebraical sum, and its line of action is parallel to that of any of them, and passes through a point whose position depends solely on the positions of the fixed points and the relative magnitude of the forces. For two of the forces are equivalent to their sum acting parallel to them and through a fixed point: this resultant and a third force of the system are also equivalent to the algebraical sum of the three acting parallel to them through a fixed point; in this way we can go on reducing the number of the forces until we arrive at the final resultant acting through a fixed point. We shall necessarily arrive at the same fixed point, whatever be the order in which we compound the forces: for if by compounding them in different orders we obtain two points, at either of which the resultant may act, its line of action must be the line joining the two points (Art. 53), which is inconsistent with its being always parallel to the directions of the original forces.

It is assumed above that the algebraical sum of the forces is not zero, otherwise, if they are not in equilibrium, they will not reduce to a single force, but to a couple.

83. *Def.* The *Centre* of a number of *parallel forces* acting at fixed points, is the point at which their resultant always acts, however their direction alters, so long as their relative magnitudes remain the same.

If the points at which the parallel forces act lie in *one* plane, we can find an expression for the distance of the centre of the forces from any straight line in the plane.

Let A_1, A_2, A_3, &c. be the points of application of the parallel forces, P_1, P_2, P_3, &c., and let C be their centre.

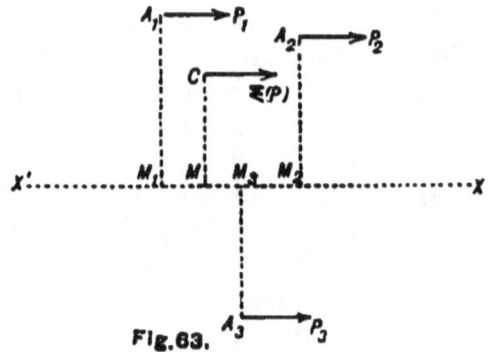

Fig. 63.

Let $X'X$ be any straight line in the plane containing the points of application. Draw A_1M_1, A_2M_2, &c., CM perpendicular to $X'X$. Let x_1, x_2......$\bar{x}$ be the lengths of these perpendiculars, which are reckoned positive or negative according to the side of the line on which the corresponding point of application lies.

As the position of C is independent of the *direction* of the forces, it will not be affected by supposing P_1, P_2, &c. to act parallel to $X'X$. Since the algebraical sum of P_1, P_2, &c. acting at C, is the resultant of these parallel forces, the algebraical sum of the moments of P_1, P_2, &c. about any point in $X'X$ is equal to the moment of their algebraical sum at C, about the same point in $X'X$.

$$\therefore P_1 x_1 + P_2 x_2 + \ldots = (P_1 + P_2 + \ldots)\bar{x};$$

$$\therefore \bar{x} = \frac{P_1 x_1 + P_2 x_2 + \ldots}{P_1 + P_2 + \ldots} = \frac{\Sigma(Px)}{\Sigma(P)}.$$

As we can find in this way the distance of the centre of a number of parallel forces acting at fixed points in one plane, from *two* intersecting straight lines in that plane, its position is completely determined.

84*. When the points of application of the parallel forces are not in one plane, we can find an expression for the distance of the centre from any given *plane*.

Let $A_1, A_2,$ &c. be the points at which the forces $P_1, P_2,$ &c. respectively act: let C be the centre. Draw $A_1 M_1, A_2 M_2,$

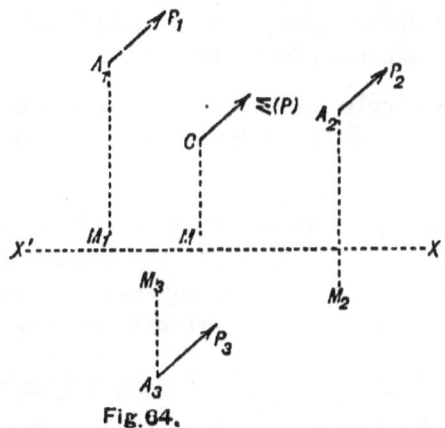

Fig. 64.

&c., CM perpendicular to the given plane; let $x_1, x_2 \ldots \bar{x}$ be these respective distances, which are reckoned positive or negative according to which side of the plane the corresponding points lie. Let $X'X$ be any straight line in the plane.

Since the position of C is independent of the direction of the forces, it will not be affected by supposing this direction to be parallel to the plane and at right angles to $X'X$.

As the resultant of P_1, P_2, &c. is their algebraical sum acting at C, the algebraical sum of the moments of P_1, P_2, &c. about $X'X$ is equal to the moment of their algebraical sum acting at C about $X'X$.

$$\therefore P_1 . A_1M_1 + P_2 . A_2M_2 + \&c. = (P_1 + P_2 + ...) CM;$$
$$\therefore P_1x_1 + P_2x_2 + \&c. = (P_1 + P_2 + ...) \bar{x};$$
$$\therefore \bar{x} = \frac{P_1x_1 + P_2x_2 + ...}{P_1 + P_2 + ...} = \frac{\Sigma(Px)}{\Sigma(P)}.$$

When we have found the distance of C from *three* planes which have only one point in common, its position is completely determined.

Ex. 1. O is the intersection of the diagonals of a square $ABCD$, whose side is 1 foot long: find the position of the centre of like parallel forces acting at A, B, C, D and O, respectively proportional to 4, 3, 4, 6 and 9.

Ans. In DB, distant $5\tfrac{1}{3}$ inches from AD.

Ex. 2. At the angular points A, B, C of an equilateral triangle, like parallel forces of 1, 2, and 3 lbs. respectively act: find the distance of their centre from C. *Ans.* $\tfrac{1}{6}\sqrt{7}.AB$.

Ex. 3. At four of the angles of a regular hexagon taken in order, parallel forces proportional to 3, -2, 7 and -5 act: find the magnitude of the forces that must act at the remaining angles, in order that the centre of the six parallel forces may be the centre of the hexagon. *Ans.* 6, -1.

85. *Def.* Let A_1, A_2, A_3, &c. be a number of particles of masses m_1, m_2, m_3, &c. respectively; then if a point C_1 be taken in A_1A_2, so that

$$m_1 . C_1A_1 = m_2 . C_1A_2,$$

this point is called the *Centre of Mass* or the *Centre of Inertia* of the particles A_1 and A_2. The centre of mass of A_3 and a particle of mass $(m_1 + m_2)$ situate at C_1 is the centre of mass of A_1, A_2 and A_3. That of A_4 and a particle of mass $(m_1 + m_2 + m_3)$ situate at the centre of mass of

A_1, A_2 and A_3 is the centre of mass of A_1, A_2, A_3 and A_4. Continuing this process we obtain the centre of mass of any number of particles.

From this definition of the centre of mass of a number of particles it is clear that its position is the same as that of the centre of a number of like parallel forces acting one on each of the particles, each force being proportional to the mass of the particle on which it acts. Hence if the particles of masses m_1, m_2... be at distances x_1, x_2... respectively from a given plane, the distance of their centre of mass from that plane is $\Sigma\,(mx)/\Sigma\,(m)$; or, in other words, the distance from a given plane of the centre of mass of a number of particles is obtained by multiplying the mass of each by its distance from the plane, and dividing the algebraical sum of the products by that of the masses.

If the particles are not fixed in position, but move so that the configuration formed by them is unaltered in shape, their centre of mass will be a point moving with the configuration, but occupying a position fixed relatively to it.

Def. The product of a mass into the distance of its centre of mass from any plane or line is termed its *Moment about that plane or line*.

We see from the above, that *the algebraical sum of the moments of a number of masses about any plane, and if they are coplanar, about any line, is equal to the moment of the whole mass collected at the centre of mass about the same plane or line.*

86. Let us suppose that the above system is acted on by a number of like parallel forces, one on each particle, their magnitudes being proportional to the masses of the particles on which they respectively act: now, no matter how much the direction of the forces varies, or to what extent the particles move, so long as the configuration formed by them remains the same, the resultant of these forces will always pass through the centre of mass, which is fixed

relatively to the configuration. Since the magnitude of a particle's weight is proportional to its mass and its direction is towards the earth's centre, the weights of a system of particles which are not far from one another in comparison with their distance from the earth's centre, are forces approximately parallel, and also proportional to the masses of the particles on which they act. The line of action of their resultant then will approximately always pass through a point fixed relatively to the configuration formed by the particles, if that configuration does not alter, though it move as a whole. This point, which we have called the centre of mass of the system, is on this account often called its *Centre of Gravity*.

We may define the Centre of Gravity thus: *The Centre of Gravity of a body is the point, fixed relatively to the body and through which the resultant of the weights of the particles composing it always acts, however the body move, provided it always moves as if it were rigid.*

Strictly speaking, there is not of necessity any such point for every body, because the weights of the particles composing the body are not accurately parallel, but if they are very nearly so their resultant will pass very close to the centre of mass, if it does not pass through it.

It is not assumed in the definition of the centre of gravity that the body is a rigid one: any body whatsoever, a flexible string for instance, or a mass of liquid, will have a centre of gravity corresponding to every definite shape of the body, though its position in the body will generally alter with an alteration of the body's shape.

If a body be such that the action of gravity on it can always be reduced to a single force passing through a point fixed relatively to the body, whatever be its position relatively to the earth, the body is termed a *Centrobaric Body*, and the point its *Centre of Gravity*, in a stricter sense than is usually attached to the term.

87. *Def.* When a substance is such that the mass of any volume of it is proportional to that volume, it

CENTRES OF MASS. 127

is said to be *homogeneous*, or of *uniform density:* when this is not the case, it is said to be *heterogeneous*, or of *variable density.*

When a substance is homogeneous its density is measured by the numerical measure of the mass in a unit of volume.

When the density of a substance *varies*, the *average* density of any volume is measured by the ratio of the numerical measure of its mass to that of its volume. The density *at any point* is measured by the limit of the average density of an indefinitely small volume containing the point in question.

88. Prop. *Having given the centres of mass of a body and of one part of it, to find that of the remaining part.*

Let m_1, m_2 be the masses of the two parts forming the body, C_1, C_2 their respective centres of mass. Join $C_1 C_2$,

$$C_1 \quad\quad C \quad\quad\quad\quad C_2$$

Fig. 65

and take C between C_1 and C_2, such that $m_1 \cdot CC_1 = m_2 \cdot CC_2$: then C is the centre of mass of the whole.

Since C_1, C, C_2 are connected in this way, it is perfectly clear that if C_1 and C are given, C_2 is the point obtained by producing $C_1 C$ to a distance $= (m_1/m_2) \cdot CC_1$.

Cor. In a similar way we can obtain the centre of one part of a system of parallel forces when the centres of the whole system and of the remaining part are known.

89. Prop. *If the mass of each of a series of particles be multiplied by the square of its distance from any given point, the sum of the products so obtained is equal to the sum of the products obtained by multiplying the mass of each particle by the square of its distance from the centre of mass of all the particles, together with the product of the whole mass into the square of the distance of the given point from the centre of mass.*

Let A_1, A_2, &c., A_n be n particles of mass $m_1, m_2 \ldots m_n$; let G be their centre of mass, and O any point whatsoever.

Join GO, and draw A_1M_1, A_2M_2, &c. perpendicular to GO.

Then $AO^2 = AG^2 + OG^2 - 2OG \cdot GM.$

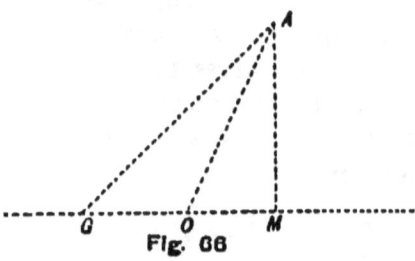

Fig. 66

The $-$ sign in this equation refers to the above figure where M and O are on the same side of G, but if we agree that GM shall be reckoned positive when M is on the same side of G as O, and negative when on the other, the equation holds for all figures.

Hence

$$\Sigma(m \cdot AO^2) = \Sigma(m \cdot AG^2) + \Sigma(m \cdot OG^2) - 2\Sigma(m \cdot OG \cdot GM)$$
$$= \Sigma(m \cdot AG^2) + OG^2 \cdot \Sigma(m) - 2OG \cdot \Sigma(m \cdot GM).$$

But since G is the centre of mass of the n particles, $\Sigma(m \cdot GM)$ is zero (Art. 85), and we have

$$\Sigma(m \cdot AO^2) = \Sigma(m \cdot AG^2) + OG^2 \cdot \Sigma(m).$$

(In the above proof, it is not assumed that A_1, A_2, &c. are in one plane.)

Cor. If the mass of each of a series of particles be multiplied by the square of its distance from any given point, the product so obtained is least when the given point is the centre of mass of the system of particles.

90. We shall now investigate the positions of the centres of mass of some of the simpler geometric figures.

CENTRES OF MASS.

Prop. *If a body consists entirely of pairs of particles, such that those forming each pair are of equal mass and at equal distances from, but on opposite sides of, a certain point, that point is the centre of mass of the body.*

For this point is clearly the centre of mass of each pair, and therefore of all the pairs, i.e. of the whole body.

Hence the centre of mass of a thin rod, uniform in density and sectional area, is its middle point: that of a lamina, uniform in thickness and density, and in shape, a circle, ellipse, or parallelogram, is its centre of figure. Also the centre of figure of a homogeneous sphere, ellipsoid, or parallelepiped is its centre of mass. The centres of mass of many other figures can be thus determined.

91. Prop. *If a body consists entirely of pairs of particles, those forming each pair being of equal mass and such that the middle point of the line joining them is on a certain straight line or plane, the centre of mass of the body lies in that straight line or plane.*

For this straight line or plane contains the centre of mass of every pair of particles and therefore that of the whole body.

Hence any straight line or plane which divides a homogeneous body symmetrically, contains its centre of mass. For instance the centre of mass of the volume of surface of a right circular cone, with its base at right angles to its axis, lies in the axis: that of a segment of an ellipse or parabola lies in the diameter conjugate to the chord cutting off the segment.

When we speak of the centre of mass of a surface or plane figure, we suppose the figure to be of very small uniform thickness. Similarly a line or curve is supposed to be of very small uniform sectional area.

92. *To find the centre of mass of a plane triangle.*

Let ABC be the triangle. Bisect BC in D, and join AD. Draw bdc parallel to BC, meeting AD in d.

Then $bd : BD = Ad : AD = dc : DC$;

$$\therefore bd = dc.$$

Similarly it may be shewn that AD bisects any other line parallel to BC. Hence the triangle consists entirely of

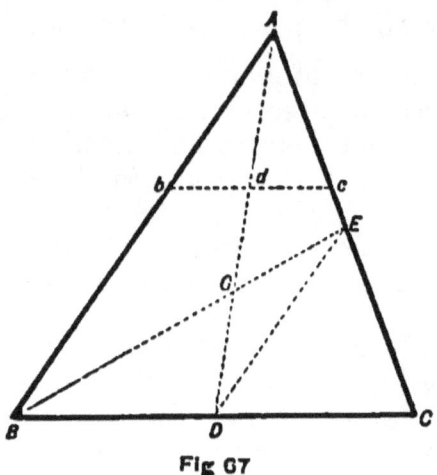

Fig 67

pairs of particles, those forming each pair being of equal mass, and such that AD bisects the line joining them: the centre of mass of the triangle is therefore in AD. (Art. 91.)

Bisect AC in E, and join BE meeting AD in G.

Then we can prove as before that the centre of mass of the triangle lies in BE, as well as in AD: it must therefore be G, their point of intersection.

Join DE: since D, E are the middle points of BC, AC respectively, DE is parallel to AB, and

$$DE = \tfrac{1}{2}AB;$$
$$\therefore AG : GD = AB : DE = 2 : 1;$$
$$\therefore AG = \tfrac{2}{3}AD.$$

Hence the centre of mass of a triangle is obtained by joining the vertex with the middle point of the base and taking the point two-thirds the way down this line from the vertex.

CENTRES OF MASS.

The centre of mass of the triangle ABC coincides with that of three equal particles placed at A, B, and C. For the centre of mass of those at B and C is at D, half-way between them: and that of all three will be in AD at a point g, such that

$$Dg : gA = 1 : 2, \text{ or } Dg = \tfrac{1}{3}DA.$$

Hence G and g are the same point.

Cor. By drawing an indefinitely large number of lines parallel to BC and at equal distances from one another, the triangle ABC may be divided into an infinitely large number of indefinitely narrow strips, of equal breadth, and having their centres of mass in AD. Now the mass of each strip is proportional to its area, i.e. to its length, and therefore to its distance from A measured along AD: also for the purpose of finding the centre of mass of the whole we may suppose the mass of each portion collected at its centre of mass. The problem therefore of finding the centre of mass of all these strips, i.e. of the triangle, is the same as that of finding the centre of mass of an infinite number of masses arranged along AD at equal, but indefinitely small distances, each mass being proportional to its distance from A. The centre of mass in the latter case, then, is at a distance from A equal to two-thirds of AD. Hence the centre of mass of a thin rod, of uniform sectional area, but such that the density at any point varies as its distance from one end, is distant from that end two-thirds the length of the rod.

Also the centre of mass of the portion of a paraboloid of revolution, cut off by a plane perpendicular to the axis, is at a distance from the vertex equal to two-thirds the length of the axis cut off.

For let the paraboloid be divided into an indefinitely large number of thin slices of equal thickness by planes perpendicular to the axis. Then the volume of any slice is proportional ultimately to the square of its radius, i.e. to its distance from the vertex, whence the result given above follows at once by the preceding corollary.

Ex. 1. Weights of 1, 4, 2, 3 lbs. are placed at the corners taken in order of a parallelogram $ABCD$; a weight of 10 lbs. is also placed at O, the intersection of diagonals; find the position of their centre of mass.

Ans. If E be the middle point of BC, the point required is in OE, at a distance from O equal to one-tenth of OE.

Ex. 2. A line AB is bisected in C_1, C_1B in C_2, C_2B in C_3, and so on *ad infinitum*, and weights equal to P, $\dfrac{P}{2}$, $\dfrac{P}{2^2}$, &c. are placed at the points C_1, C_2, C_3, &c. Prove that the distance of the centre of mass of the whole system from B is equal to one-third of AB.

Ex. 3. Find the centre of mass of seven equal particles placed at the angular points of a regular octagon.

Ans. If A be the unoccupied angular point and O the centre of the octagon, the required point is in AO produced, at a distance from O equal to $\frac{1}{7}AO$.

Ex. 4. A square $ABCD$ is divided into four equal triangles, by its diagonals, which intersect in O: if the triangle OAB be removed, find G, the centre of mass of the remaining three. Prove that if E be the middle point of CD, G is in OE, and $OG = \frac{1}{9}AB$.

Ex. 5. The sides of a square $ABCD$ are bisected, and the points of bisection of the opposite sides joined. If the small square, having the angle A, be removed, find G the centre of mass of the remaining three.

Ans. G is in AC and $CG = \frac{5}{14}AC$.

Ex. 6. Out of a circular lamina of radius r is cut a circle, whose diameter coincides with a radius of the lamina: find the position of the centre of mass of the remainder.

Ans. The c. m. is at a distance from the centre of the lamina equal to $\frac{1}{6}r$.

Ex. 7. A figure consists of a square and an isosceles triangle, whose base is one of the sides of the square: if the side of the square be 6 inches, and the height of the triangle be 6 inches, find the centre of mass of the figure.

Ans. Within the square, $\frac{1}{2}$ of an inch from the base of the triangle.

CENTRES OF MASS. 133

Ex. 8. A uniform rod, 18 inches long, is bent so that the two parts, 8 and 10 inches long respectively, are at right angles to one another. Find the distance between the centres of mass of the new shape and the original. *Ans.* $\frac{10}{9}\sqrt{2}$ inches.

Ex. 9. Equal weights are placed at $n-2$ of the corners of a regular n-sided polygon: find their centre of mass.

Ans. If A, B be the unoccupied corners, C the middle point of AB, and O the centre of the polygon, the centre of mass is in CO produced at a distance from O equal to $\dfrac{2}{n-2} OC$.

Ex. 10. Having given the position of the centre of mass of two particles A and B, and also that of A and C, find that of B and C.

Ans. Join B with E, the c.m. of A and C, and C with D, the c.m. of A and B; let these two lines meet in G. The point where AG meets BC is the c.m. of B and C.

Ex. 11. Assuming that the pressure on an indefinitely small area below the surface of a liquid is perpendicular to the area and varies as the area and its depth below the surface conjointly: find where the resultant pressure on a parallelogram, one of whose sides is in the surface of the liquid, acts.

Ans. At a point whose depth below the surface is two-thirds that of the lowest side.

Ex. 12. With the same assumption as in the last example, shew that the resultant pressure on any plane area below the surface of a liquid is proportional to the area and the depth of its centre of mass below the surface conjointly.

Ex. 13. Find the centre of mass of a quadrilateral, two of whose sides are parallel to one another, and respectively 6 inches and 14 inches long, while the other sides are each 8 inches long.

Ans. In the line joining the middle points of the two parallel sides, at a distance of $\frac{28}{15}\sqrt{3}$ inches from the longer side.

Ex. 14. Find also the centre of mass of the perimeter of the above quadrilateral.

Ans. In the line joining the middle points of the parallel sides, at a distance from the greater equal to $\frac{14}{9}\sqrt{3}$ inches.

93. *To find the centre of mass of a triangular pyramid.*

Let $ABCD$ be the pyramid. Bisect BC in E, join AE, and take H in it, so that
$$AH = \tfrac{2}{3}AE.$$

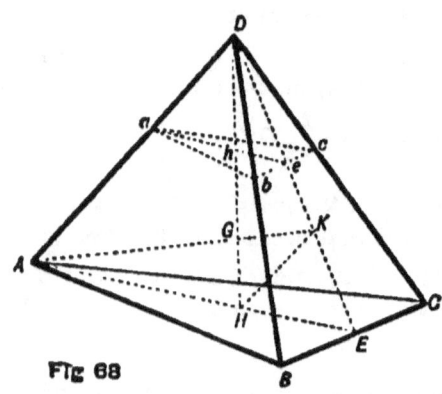

Fig 68

Let abc be a section of the pyramid, made by a plane parallel to ABC, and let ae be its intersection with the plane ADE.

Since the planes ABC, abc are parallel, bc, BC are also parallel;
$$\therefore be : ec = BE : EC;$$
$$\therefore be = ec,$$
and e is the middle point of bc.

Similarly ae, AE are parallel, and
$$ah : ae = AH : AE = 2 : 3,$$
i.e. h is the centre of mass of the triangle abc.

Hence, if we suppose the pyramid divided into an infinitely large number of indefinitely thin triangular slices made by planes parallel to ABC, the centre of mass of each slice will lie in the line DH, which must therefore contain the centre of mass of the pyramid. Join DE, and take K in it so that $DK = \tfrac{2}{3}DE$; join AK intersecting DH in G. Then, as before, we can shew that the centre of mass of the pyramid lies in AK, as well as in DH; the point of intersection G of these two lines must therefore be the required centre of mass. Join HK.

$$AH = \tfrac{2}{3}AE, \text{ and } DK = \tfrac{2}{3}DE;$$
$$\therefore HK \text{ is parallel to } AD,$$
and $\quad DG : GH = AD : HK = AE : HE = 3 : 1,$
$$\therefore DG = \tfrac{3}{4}DH.$$

Hence the centre of mass of the pyramid is in the line drawn from any vertex to the centre of mass of the opposite face, and is such that its distance from the former point is three times its distance from the latter.

Cor. *The centre of mass of a triangular pyramid coincides with that of four particles of equal mass placed at its angular points.*

For the centre of mass of the particles at A, B and C is H, and therefore that of the four is in HD, and at a distance from D equal to three times its distance from H; it is therefore G, the centre of mass of the pyramid.

94. If the above pyramid be divided into an indefinitely large number of indefinitely thin slices, such as abc, of the same thickness, we may suppose the mass of each slice to be collected at its centre of mass h, which lies in DH: also the mass of any slice abc is proportional to its area, since they are of equal thickness, and therefore to the square on Dh. Hence finding the centre of mass of a triangular pyramid is the same problem as finding that of an indefinitely large number of masses arranged at equal but indefinitely small intervals along a straight line, each mass being proportional to the square of its distance from one end of the line. We infer then that the centre of mass in the latter case is at a distance from this end equal to three-quarters the length of the line. For instance, the centre of mass of a thin rod of uniform thickness, but whose density varies as the square of the distance from one end, is the point whose distance from this end is three-quarters the length of the rod.

136   STATICS.

95*. *To find the centre of mass of a pyramid having any given rectilinear plane figure for its base.*

Let V be the vertex of the pyramid, $ABCDE$ the perimeter of its base.

Let $abcde$ be a section of the pyramid made by a plane parallel to the base. Let PQR be any straight

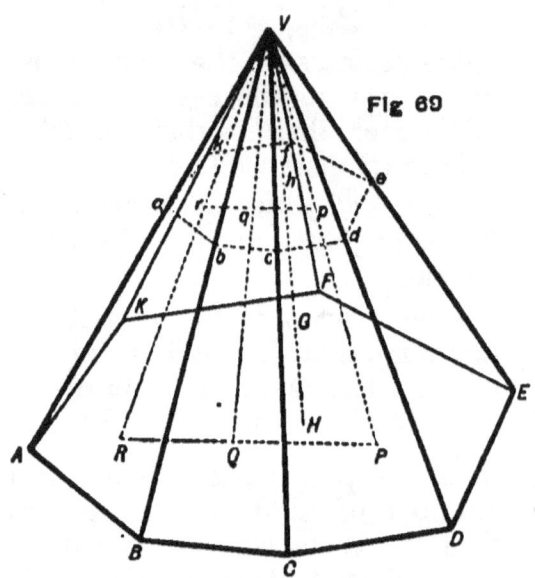

Fig 69

line in the plane of the base; join VP, VQ, VR, cutting the plane $abcd$ in p, q, r respectively; p, q, r may be said to be the *corresponding points* to P, Q, R respectively. Since pqr, PQR are the sections of parallel planes made by the plane PVQ, they are parallel;

$$\therefore PQ : pq = VQ : Vq = QR : qr,$$
$$\therefore pq : qr = PQ : QR.$$

Hence, if Q be the centre of mass of two given particles at P and R, q will be that of particles at p and r, provided the masses of the latter particles have the same ratio to one another as those of the former have. Similarly, if we have any number of particles at different points of the base and also another set of particles at the corresponding

CENTRES OF MASS. 137

points of the parallel section, the mass of each particle of one set bearing a constant ratio to that of the corresponding particle of the other set, we could shew in the same way as we have done for two, that the centres of mass of the two sets are corresponding points, i.e. that they both lie in a straight line passing through the vertex. But we may suppose the two sections $ABCD$, $abcd$ to be made up each of a number of equal particles, the positions of the particles forming one set corresponding to the positions of those forming the other set. Hence, if H be the centre of mass of the base, the point h, where VH cuts the section $abcd$, is the centre of mass of the latter. Dividing then the pyramid up into an infinitely large number of indefinitely thin slices cut off by planes parallel to the base, we see that the centre of mass of each slice and therefore that of the whole pyramid lies in VH. But the pyramid may be divided into a number of triangular pyramids $VHAB$, $VHBC$, &c. and the centre of mass of each of these will lie in a plane parallel to the base, and at a distance from it one quarter the distance of the vertex from it. The centre of mass of the pyramid must therefore be at G, the point where this plane cuts VH; i.e. the centre of mass is found by joining the vertex with the centre of mass of the base, and taking a point in the joining line at a distance from the former point three times its distance from the latter.

Cor. Since a cone or pyramid with a curvilinear base may be regarded as the limiting case of a pyramid with a rectilinear base, when the number of sides is indefinitely large, we can find the centre of mass of a cone in exactly the same way as we find that of a pyramid with a rectilinear base.

96*. *The centre of mass of the surface of a pyramid with a rectilinear base.*

If the pyramid be the one in fig. 69, its surface may be divided into a number of triangles having the common vertex V: the centre of mass of each triangle and therefore that of the whole surface will lie in a plane parallel to the base and at a distance from the vertex two-thirds that of the base.

Cor. As a cone is the limiting case of a pyramid, when the number of sides of the base is indefinitely increased, the centre of mass of the surface of a cone will also lie in a plane parallel to the base and at a distance from the vertex two-thirds that of the base.

97. *To find the centre of mass of an arc of a circle.*

Let ABC be the arc, subtending an angle 2α at the centre O.

Draw OB bisecting the angle AOC: it is clear from the principle of symmetry of Art. 91 that the centre

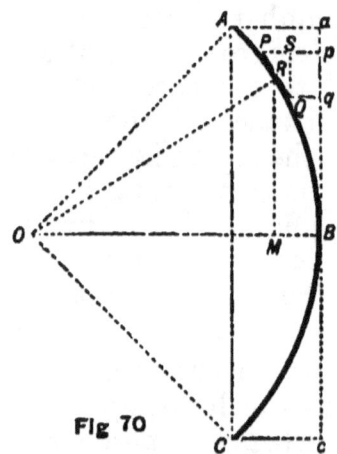

Fig 70

of mass is in OB. Construct a regular polygon circumscribing the arc; let PQ be one side of it, touching the circle at R. Draw Aa, Pp, Qq, Cc perpendicular to the tangent aBc at B, RM perpendicular to OB, and QS to Pp.

The right-angled triangles PSQ, ORM are similar, since QS, PQ are respectively perpendicular to OM, OR.

CENTRES OF MASS. 139

$$\therefore PQ : QS = OR : OM;$$

$$\therefore PQ \cdot OM = OR \cdot QS = OB \cdot pq;$$

$$\therefore \Sigma (PQ \cdot OM) = \Sigma (OB \cdot pq) = OB \cdot \Sigma (pq);$$

$$\therefore OG \cdot \text{perimeter of polygon} = OB \cdot ac = OB \cdot \text{chord } AC$$
$$\text{(Art. 85)};$$

where G is the centre of mass of the polygon.

Also, when the sides of the polygon are taken indefinitely small, the limit of the perimeter of the polygon is that of the arc, and their centres of mass are also coincident. Hence the distance of the centre of mass of the arc ABC from O is

$$\frac{\text{radius } OB \cdot \text{chord } AC}{\text{arc } ABC} = \frac{r \sin \alpha}{\alpha}.$$

The centre of mass of a uniform circular arc may also be found as follows:

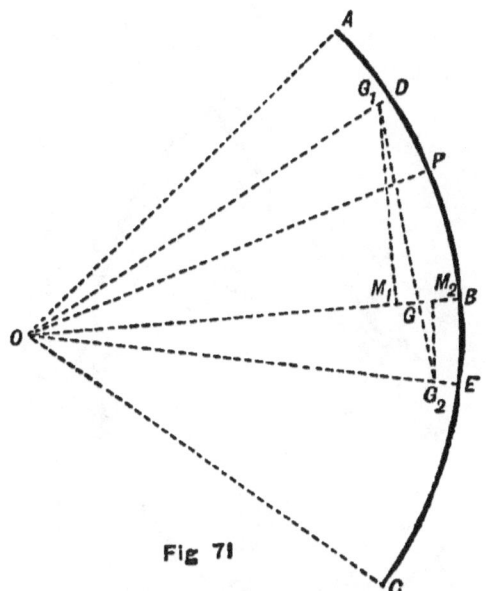

Fig 71

Let AC be any arc (2α), O the centre. Draw OP dividing the arc AC into any two unequal parts AP (2θ) and PC (2ϕ).

Draw OB, OD, OE bisecting the arcs AC, AP, PC respectively.
$$\angle DOB = \angle AOB - \angle AOD = \alpha - \theta = \phi.$$
Similarly $\angle EOB = \theta$.

G, the C.M. of the arc AC, is (Art. 91) in OB, G_1 that of AP in OD, and G_2 that of PC in OE.

Draw G_1M_1, G_2M_2 perpendicular to OB.
Then
$$GG_1 \cdot \text{arc } PA = GG_2 \cdot \text{arc } PC,$$
$$\therefore \theta \cdot G_1M_1 = \phi \cdot G_2M_2,$$
$$\therefore \theta \cdot OG_1 \sin DOB = \phi \cdot OG_2 \sin EOB,$$
$$\therefore \frac{OG_1 \cdot \theta}{\sin \theta} = \frac{OG_2 \cdot \phi}{\sin \phi},$$

i.e. $\dfrac{OG \cdot \theta}{\sin \theta}$ is independent of θ, and is therefore equal to its value when θ vanishes. But when θ vanishes $\theta / \sin \theta = 1$, and OG_1 is clearly r.

Hence $OG_1 = \dfrac{r \sin \theta}{\theta}$ and $OG = \dfrac{r \sin \alpha}{\alpha}$.

98*. *To find the centre of mass of the sector of a circle.*

Let ABC be the sector: from O, the centre of the

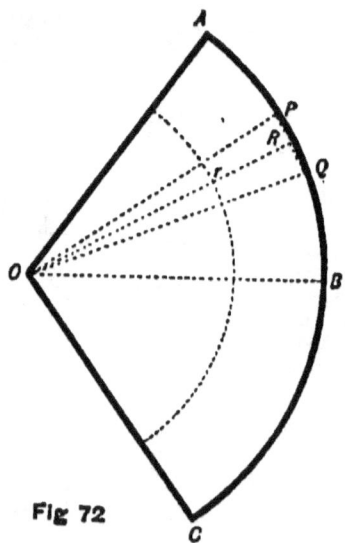

Fig 72

circle, draw OB bisecting the angle AOC. Then (Art. 91)

the centre of mass of the sector is clearly in OB. In the arc ABC inscribe a regular polygon; let PQ be one of its sides, R the middle point of PQ: join OR, and take r in OR such that Or is equal to $\frac{2}{3}OR$. Then the centre of mass of the triangle OPQ is r, and the centres of mass of all such triangles are arranged at equal angular intervals along the arc of a circle of radius Or, and whose angle is AOC. But when the number of the sides of the polygon is increased indefinitely, Or becomes equal to $\frac{2}{3}OB$ ultimately, the sum of the triangles of which OPQ is a type becomes the sector AOC, and the centre of mass of the latter is that of an infinite number of equal masses arranged at equal angular intervals along an arc of a circle of radius $\frac{2}{3}OB$, and whose angle is equal to AOC. But the centre of mass of the masses arranged along this arc is that of the arc itself. Therefore the distance of the centre of mass of the sector from O

$$= \frac{2}{3} \cdot \frac{OB \times \text{chord } AC}{\text{arc } AC}.$$

Cor. As the segment of a circle is the difference between a sector and a triangle, its centre of mass can be found by the method of Art. 88.

The centre of mass of the portion of a circle cut off by two parallel lines can also be obtained, since the figure consists of the difference of two segments.

99*. *To find the centre of mass of the belt of a sphere cut off by two parallel planes.*

Let AB be the arc of a circle, which by revolving about OE generates the belt $ABCD$ of a sphere in question. Then (Art. 91) the centre of mass of the belt lies in OE.

Let PQ be the side of a regular polygon, circumscribing the arc AB; let R be the middle point of PQ, where it touches the circle. Produce PQ to meet OE in T, and draw PM, RK, QN perpendicular to OE, and QL perpendicular to PM.

The area of the frustum of the cone, generated by the revolution of PQ about OE

$$= PT \cdot \pi PM - QT \cdot \pi QN$$
$$= \pi (PR + RT) PM - \pi (RT - RQ) \cdot QN$$
$$= \pi RT \cdot PL + 2\pi PR \cdot RK$$
$$= 2\pi PQ \cdot RK, \text{ by similar triangles } PQL, RKT$$
$$= 2\pi OR \cdot MN \text{ by similar triangles } PQL, ORK$$

= the area of the belt cut off by the planes PM, QN from the cylinder circumscribing the sphere and having its axis along OE.

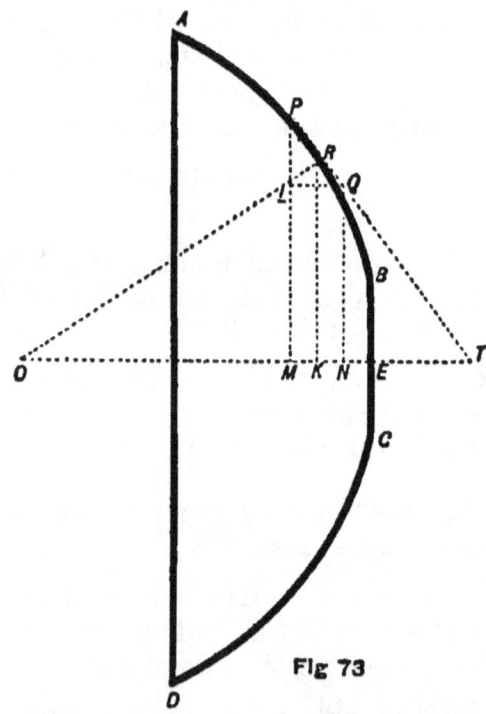

Fig 73

Hence the sum of the areas of any number of frusta of cones, of which the one considered is a type, is equal to the sum of the areas of the corresponding belts of the circumscribing cylinder. But ultimately, when the number of the sides of the circumscribing polygon is taken indefinitely large, the sum of the areas of the frusta of the cones becomes the area of the belt of the sphere. Hence the area of the belt of a sphere, cut off by parallel planes, is equal to that of the coaxial circumscribing cylinder cut off by the same planes.

Let G be the centre of mass of the belt of $ABCD$ of the sphere, G' of the corresponding belt of the cylinder. Then

OG . area of $ABCD =$ moment of $ABCD$ about the plane through O perpendicular to OE

$\quad = \Sigma\,(2\pi\,OA\,.\,MN\,.\,OK)\qquad$ (Art. 85)

$\quad =$ moment about the same plane of the belt corresponding to $ABCD$ of the cylinder

$\quad = OG'$. area of the belt of the cylinder;

$\therefore\ OG = OG'.$

Therefore G, G' are coincident, i.e. G is in OE, half-way between the planes which cut off the belt. (Art. 90.)

100*. *The centre of mass of the volume of a sector of a sphere.*

Let OAC be the spherical sector generated by the revolution of the circular sector AOB about OB. The centre of mass is in OB. (Art. 91.)

Imagine the sector to be divided into an infinite number of indefinitely small pyramids having the common vertex O. The centre of mass of each of these pyramids will lie on a spherical cap abc, generated by the revolution of ab, the arc of a circle, of radius three-quarters that of ABC, and the same vertical angle AOB. Supposing the mass of each pyramid to be collected at its centre of mass, the centre of mass of the sector AOC is clearly the same as that of the spherical cap abc: its distance from O therefore is equal to $\frac{1}{2}(Om + Ob)$ or $\frac{3}{8}(OM + OB)$.

If the sector be a hemisphere, OM vanishes, and the distance from the centre of the centre of mass of the volume of the hemisphere is $\tfrac{3}{8}$ of the radius.

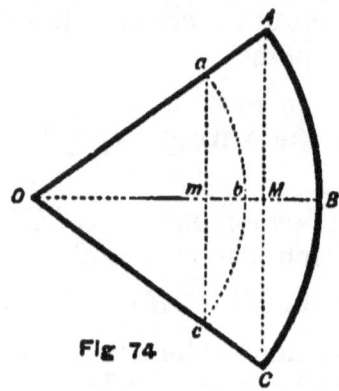

Fig 74

Cor. As a spherical segment, the solid figure cut off a sphere by a plane, is the difference between a spherical sector and a right circular cone, its centre of mass can be found by the method of Art. 88.

The centre of mass of the solid cut off a sphere by two parallel planes can also be obtained, since the figure is the difference of two spherical segments.

Ex. 1. With the same assumption as in Ex. 11, p. 133, find where the resultant pressure acts on a triangle whose vertex is in the surface of a liquid and whose base is parallel to the surface, but below it.

Ans. At a point whose depth below the surface is three-quarters that of the base.

Ex. 2. Find the centre of mass of a segment of a circle.

Ans. It is in the diameter bisecting the segment, at a distance from the centre $\tfrac{4}{3} r \sin^3 a / (2a - \sin 2a)$, where r is the radius, and $2a$ the angle the segment subtends at the centre.

Ex. 3. If a figure consist of a cone and a hemisphere on the same base, find the height of the cone in order that the centre of mass of the whole may be the centre of the hemisphere.

Ans. $\sqrt{3}$ times the radius of the hemisphere.

Ex. 4. Find the position of the centre of mass of a frustum of a cone, when the radii of the faces are 4 inches and 8 inches respectively, and the distance between them 7 inches.

Ans. In the line joining the centres of the faces at a distance of $4\frac{1}{4}$ inches from that of the smaller face.

Ex. 5. Find also the position of the centre of mass of the surface of the above figure.

Ans. At a distance of $3\frac{3}{6}$ inches from the centre of the smaller face.

Ex. 6. From a cube is cut a tetrahedron, three of whose edges are the edges of the cube which meet in one of the corners. Find the centre of mass of the remainder.

Ans. In the diagonal of the cube through the corner from which the tetrahedron is cut off, and at a distance from that corner equal to $\frac{1}{18}$ of the diagonal.

Ex. 7. Find the centre of mass of a segment of a sphere.

Ans. In the diameter of the sphere at right angles to the base of the segment and at a distance from the centre equal to $\frac{3}{4}(r+h)^2/(2r+h)$, where r is the radius of the sphere and h the distance of the base from the centre.

Ex. 8. A semicircular wire of uniform thickness consists of two parts AP, PB whose densities are proportional to BN, AN respectively, where PN is the perpendicular from P on AB: prove that the centre of mass of the whole is in the radius through P.

101*. *The centre of mass of a segment of a parabola.*

Let BAB' be the segment, AC being the diameter conjugate to the base BB'.

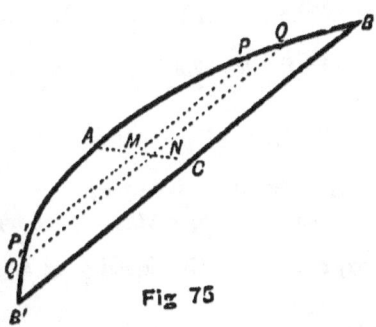

Fig 75

Divide AC into an infinite number n of indefinitely small equal parts of which MN is a typical one, the rth. Draw PMP', QNQ' chords parallel to BB'.

Let S be the focus; then
$$PM^2 = 4AS \cdot AM.$$
The centre of mass of the strip $PP'Q'Q$ is in MN. (Art. 91.)

The area $PP'Q'Q$ lies between
$$PP' \cdot MN \sin BCA, \text{ and } QQ' \cdot MN \sin BCA,$$
and the distance of its c.m. from A lies between AM and AN; therefore the sum of the moments of all the strips about a line through A perpendicular to AC lies between

$$\Sigma (PP' \cdot MN \cdot AM \sin BCA), \text{ and } \Sigma (QQ' \cdot MN \cdot AM \sin BCA),$$

i.e. between
$$4AS^{\frac{1}{2}} \sin BCA \cdot \Sigma (AM^{\frac{3}{2}} \cdot MN),$$
and
$$4AS^{\frac{1}{2}} \sin BCA \cdot \Sigma (AN^{\frac{3}{2}} \cdot MN).$$

But $AM = \dfrac{r}{n} AC$, $AN = \dfrac{r+1}{n} AC$, and $MN = \dfrac{AC}{n}$; therefore the moment of the mass of the parabola lies between

$$4AS^{\frac{1}{2}} \cdot AC^{\frac{5}{2}} \sin BCA \cdot \frac{1^{\frac{3}{2}} + 2^{\frac{3}{2}} + \ldots (n-1)^{\frac{3}{2}}}{n^{\frac{5}{2}}},$$

and
$$4AS^{\frac{1}{2}} \cdot AC^{\frac{5}{2}} \sin BCA \cdot \frac{1^{\frac{3}{2}} + 2^{\frac{3}{2}} + \ldots n^{\frac{3}{2}}}{n^{\frac{5}{2}}},$$

i.e.,
$$= \frac{4AS^{\frac{1}{2}} \cdot AC^{\frac{5}{2}} \sin BCA}{\frac{5}{2}}. \quad \text{(Appendix.)}$$

Similarly it can be shewn that the area of the segment
$$= \frac{4AS^{\frac{1}{2}} \cdot AC^{\frac{3}{2}} \sin BCA}{\frac{3}{2}}.$$

Hence the distance of the c.m. of the parabola from A
$$= \frac{8AS^{\frac{1}{2}} \cdot AC^{\frac{5}{2}} \sin BCA}{5} \div \frac{8AS^{\frac{1}{2}} \cdot AC^{\frac{3}{2}} \sin BCA}{3} = \tfrac{3}{5} AC.$$

Also the c.m. is in AC. (Art. 91.)

102*. *The centre of mass of a rod of uniform thickness, and whose density varies as the mth power of the distance from one end.*

Let AB be the rod, such that the density at any point P varies as $(AP)^m$.

CENTRES OF MASS. 147

Divide the rod into an infinite number n of indefinitely small equal parts, of which PQ is a typical one, the rth.

Fig 76

The mass of PQ lies between
$$\kappa \cdot PQ \cdot AP^m \text{ and } \kappa \cdot PQ \cdot AQ^m,$$
where κ is a constant; therefore the moment of the whole rod about A lies between
$$\Sigma\,(\kappa PQ \cdot AP^m \cdot AP) \text{ and } \Sigma\,(\kappa PQ \cdot AQ^m \cdot AQ),$$
i.e. between
$$\kappa\,(AB)^{m+2} \cdot \frac{1^{m+1} + 2^{m+1} + \ldots (n-1)^{m+1}}{n^{m+2}},$$
and
$$\kappa AB^{m+2} \cdot \frac{1^{m+1} + 2^{m+1} + \ldots n^{m+1}}{n^{m+2}},$$
i.e. $= \dfrac{\kappa AB^{m+2}}{m+2}.$

Similarly it can be shewn that the mass of the rod $= \dfrac{\kappa AB^{m+1}}{m+1}$. Hence the distance from A of the c.m.
$$= \frac{\kappa AB^{m+2}}{m+2} \div \frac{\kappa AB^{m+1}}{m+1} = \frac{m+1}{m+2} \cdot AB.$$

The centres of mass of a triangle, pyramid and paraboloid of revolution might have been obtained by methods similar to those employed in the last two articles.

Ex. 1. Find the c.m. of a tetrahedron $ABCD$, which is such that the density at all points in a plane parallel to BCD is the same and proportional to the distance of the plane from A.

Ans. If G is the c.m. of BCD, the required point is in AG, at a distance from $A = \frac{4}{5}AG$.

Ex. 2. Find the c.m. of a triangular lamina ABC, when the density at any point varies as its distance from BC.

Ans. The middle point of AD, where D bisects BC.

Ex. 3. Find the c.m. of a tetrahedron $ABCD$, when the density at any point is proportional to its distance from the face BCD.

Ans. In AG, at a distance from A equal to $\frac{3}{5}AG$, when G is the c.m. of the face ABC.

Ex. 4. The density of a conical shell standing on a plane horizontal base varies as the depth below the vertex: find the depth of the centre of mass. *Ans.* $\frac{3}{4}$ the height of the cone.

103. Prop. *When a body or system of bodies is in equilibrium under the action of gravity, mutual actions, and the action of one external supporting point, the centre of mass of the whole system, and the supporting point lie in a vertical line.*

For considering the equilibrium of the whole system, the only external forces acting on it are, its weight acting vertically at its centre of mass and the action of the supporting point: but these two forces cannot maintain equilibrium, unless their lines of action are the same, which will not be the case, unless the centre of mass and the supporting point are in a vertical line.

104. Prop. *If a rigid body be placed in contact with a smooth horizontal plane, it will be in equilibrium or not, according as the vertical line drawn through its centre of mass meets the horizontal plane within the base or not.*

By the base is meant the polygon, without re-entering angles, formed by joining the extreme points of the body in contact with the plane.

Let $ABCDE$ be the base, O the point where the vertical through G, the centre of mass, meets the plane.

(i) When O lies *within* the base.

It is obvious that the direction, in which the weight of the body acting along GO tends to turn it about the side AB of the base, is such that the points C, D, E, &c. would move *downwards* if the plane were not there to resist such motion. As the plane is there such motion is prevented.

The same remark applies to motion about every other side of the base. Hence the weight will not produce any motion: and the resistances of the plane on the base are *passive* forces which can only resist motion and not produce it. The body is therefore in equilibrium.

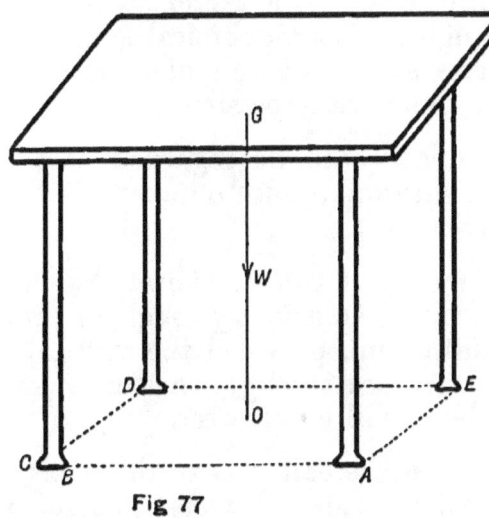

Fig 77

(ii) When O lies *outside* the base.

In this case, the base and the point O must lie on opposite sides of one or more sides of the base.

Let AB be such a side of the base.

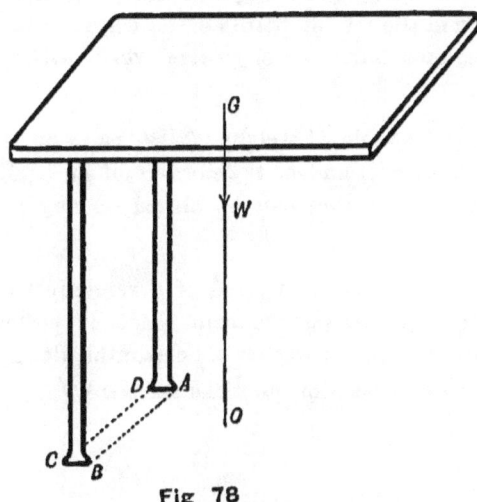

Fig 78

Now the reaction exerted by the plane on any point of the body touching it can only be vertically *upwards*, and its moment about AB is therefore of the same sign as that of the weight. The algebraical sum of the moments of all the forces about AB cannot therefore be zero, and equilibrium is therefore impossible.

If a curvilinear base be regarded as the limit of a polygonal one, with an infinite number of sides, the above reasoning applies to it.

In a similar way it can be shewn that a body placed on an *inclined* plane, sufficiently rough to prevent *sliding*, will be in equilibrium, provided the vertical through the centre of mass passes through the base, and that if it does not, the body will topple over.

It will be seen hereafter that these propositions are merely particular cases of more general propositions. (Art. 123.)

Ex. 1. A right-angled triangle ABC, whose sides AB, BC are respectively 5 and 6 feet long, is hung from the point A. Find the inclination of BC to the horizon. *Ans.* $\tan^{-1}(\frac{5}{6})$.

Ex. 2. A plane triangle is hung with its plane horizontal by three vertical chains from the middle points of its edges. How heavy must it be that a 12-stone man may walk anywhere over it without tilting it? *Ans.* 36 st.

Ex. 3. A circular table of weight 20 lbs. rests on three legs, which are on the circumference, and at the corners of an equilateral triangle. Find the greatest weight that can be placed on any part of the table without upsetting it. *Ans.* 20 lbs.

Ex. 4. A metal lamina, composed of a semicircle and an isosceles triangle (vertical angle $2a$) on the same base, is placed in a vertical plane with its curved rim resting on a horizontal plane; prove that the lamina will rest in any position provided $\tan a = \frac{1}{2}\sqrt{2}$.

CENTRES OF MASS. 151

ILLUSTRATIVE EXAMPLES.

Ex. 1. If the three diagonals of an octahedron intersect in a point O, the centre of inertia of the octahedron coincides with that of seven particles, one at O and one at each of the angular points: the mass of the particle at O being unity, and of that at each angular point the ratio of its distance from O to the diagonal through the point.

Let $ABCD$ be the plane containing two diagonals AOC, BOD: let EOF be the other diagonal. Let us find the distance of the centre of inertia of

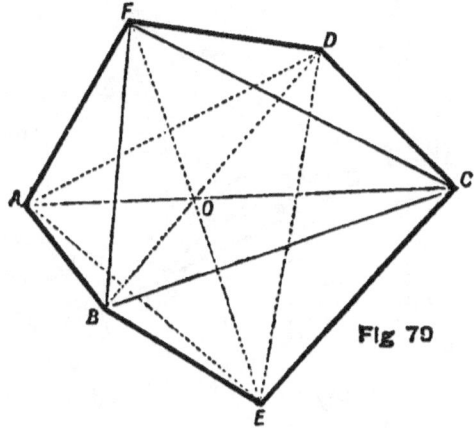

Fig 79

the octahedron from the plane $ABCD$. Let h_1 be the height of the pyramid, having base $ABCD$ and vertex F, and let h_2 be the height of the pyramid having the same base and vertex E.

$\therefore$ the volume of first pyramid : volume of second $= h_1 : h_2 = OF : OE$.

The distance of the c.i. of the octahedron from the plane $ABCD$

$$= \frac{h_1 \cdot \frac{1}{4} h_1 - h_2 \cdot \frac{1}{4} h_2}{h_1 + h_2} = \frac{1}{4}(h_1 - h_2).$$

(Here, distances from the plane $ABCD$ have been estimated positive when towards F, and negative when in the opposite direction.)

The distance from $ABCD$ of the c.i. of the seven particles

$$= h_1 \cdot \frac{OF}{EF} - h_2 \cdot \frac{OE}{EF} \div \left(1 + \frac{OF + OE}{EF} + \frac{OA + OC}{AC} + \frac{OB + OD}{BD}\right)$$

$$= \frac{h_1 \cdot \frac{h_1}{h_1 + h_2} - h_2 \cdot \frac{h_2}{h_1 + h_2}}{4} = \frac{h_1 - h_2}{4}.$$

152   STATICS.

Hence the distances of the centres of inertia of both octahedron and the seven particles from the plane $ABCD$ are the same: and it could be shewn in a similar manner that their distances from the planes $BEDF$, $ECFA$ are also the same. The two points are therefore coincident.

Ex. 2. Find the centre of mass of a segment cut off an ellipse by a straight line.

Let DPE be the segment cutting the ellipse, whose semiaxes are CA, CB. Let bpA be the auxiliary circle. Draw the ordinates DD', EE', and

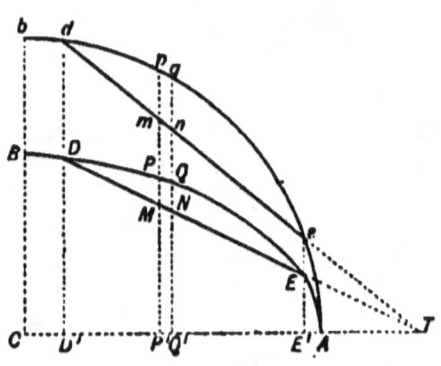

Fig 80

let them be produced to meet the circle in d, e, respectively. Join de, and produce it to meet DE in T. T is in CA produced. Draw two ordinates PP', QQ' of the ellipse indefinitely near to one another, and let them meet DE in M, N respectively. Produce them to meet de in m, n, and the circle in p, q respectively.

$$PP' : pP' = CB : CA = MP' : mP';$$
$$\therefore MP : mp = CB : CA;$$
$$\therefore \text{the area } PQNM : \text{area } pqnm = CB : CA.$$

Both elliptic and circular segments may be divided up into the same infinite number of strips, of which $PQNM$, $pqnm$ are types. Let G be the c.m. of DPE, and g that of dpe.

The distance of G from CB

$$= \frac{\text{moment of } DPE \text{ about } CB}{\text{mass of } DPE}$$

CENTRES OF MASS.   153

$$= \frac{\Sigma(PQNM \cdot CP')}{\Sigma(PQNM)} = \frac{\frac{CB}{CA}\Sigma(pqnm \cdot CP')}{\frac{CB}{CA}\Sigma(pqnm)}$$

$$= \frac{\Sigma(pqnm \cdot CP')}{\Sigma(pqnm)} = \frac{\text{moment of } dpe \text{ about } CB}{\text{mass of } dpe}$$

$$= \text{distance of } G \text{ from } CB.$$

The distance of G from CA

$$= \frac{\Sigma\left(PQNM \cdot \frac{PP'+MP'}{2}\right)}{\Sigma(PQNM)} = \frac{\frac{CB^2}{CA^2} \cdot \Sigma\left(pqnm \cdot \frac{pP'+mP'}{2}\right)}{\frac{CB}{CA} \cdot \Sigma(pqnm)}$$

$$= \frac{CB}{CA} \times \text{distance of } g \text{ from } CA.$$

But the position of g is known (Art. 98), and therefore its distances from CA and CB: hence the position of G is determined.

Ex. 3. A bowl of uniform thin material in the form of a segment of a sphere is closed by a circular lid of the same material and thickness which is hinged across a diameter. If it be placed on a smooth horizontal plane with one half of the lid turned back over the other half, shew that the plane of the lid will make with the horizontal plane an angle $\tan^{-1}\left(\frac{4}{3\pi}\tan\frac{a}{2}\right)$; a being the angle any radius of the lid subtends at the centre of the sphere of which the bowl is part.

Let EOC be the diameter about which the lid turns: BO the radius

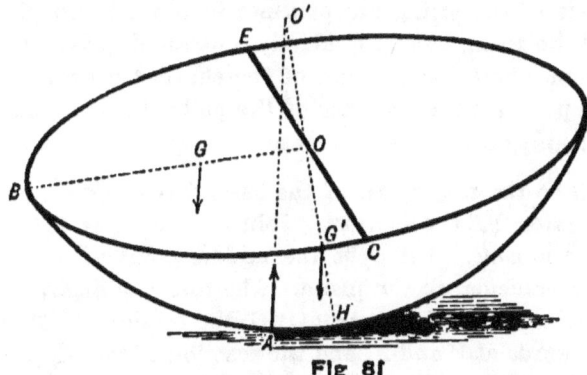

Fig 81

at right angles to it. Let O' be the centre of the sphere, and let $O'O$ meet the surface of the bowl in H. The centre of mass of the bowl is at G' in OH, such that $OG' = G'H$; that of the doubled lid is at G in OB, such that $OG = \dfrac{4OB}{3\pi}$. Draw $O'A$ vertically downwards, let θ be the angle $HO'A$, which is the inclination of the lid to the horizontal. A is the point where the bowl touches the horizontal plane. Let $r =$ the radius of the bowl. The centre of mass of the whole body must be vertically above A. (Art. 103.)

The distance of G' from $O'A = O'G' \sin\theta = (O'H - G'H) \sin\theta$

$$= \{r - \tfrac{1}{2}(r - r \cos\alpha)\} \sin\theta = \tfrac{1}{2} r (1 + \cos\alpha) \sin\theta.$$

The distance of G from $O'A = -OO' \sin\theta + OG \cos\theta$

$$= -r \cos\alpha \sin\theta + \frac{4r \sin\alpha}{3\pi} \cos\theta.$$

$\therefore 2\pi r^2 (1 - \cos\alpha) \cdot \dfrac{r}{2} (1 + \cos\alpha) \sin\theta$

$$= \pi r^2 \sin^2\alpha \left(-r \cos\alpha \sin\theta + \frac{4r \sin\alpha}{3\pi} \cos\theta \right);$$

$$\therefore \sin\theta = -\cos\alpha \sin\theta + \frac{4 \sin\alpha}{3\pi} \cos\theta;$$

$$\therefore \tan\theta = \frac{4}{3\pi} \cdot \frac{\sin\alpha}{1 + \cos\alpha} = \frac{4}{3\pi} \tan\frac{\alpha}{2}.$$

Ex. 4. A right circular cone rests with its elliptic base on a smooth horizontal table. A string fastened to the vertex and the other extremity of the longest generator passes round a smooth pulley above the cone, so that all parts of the string except those in contact with the pulley are vertical. If the string become gradually contracted by dampness and tend to lift the cone, shew that the end of the shortest generator will remain on the table provided the diameter of the pulley be less than three times the semi-axis major of the elliptic base.

Let AOB be the major axis of the base, O the centre: let VA be the longest generator, VB the shortest. Join VO, and take $VG = \tfrac{3}{4}VO$. G is the c.m. of the cone. Let C be the middle point of VA. Draw CK, GN, VM perpendicular to the plane. The forces acting on the cone are its weight W vertically downwards at G, the tensions T, T of the string vertically upwards at V and A, and the reaction of the plane on the base. We may replace the tensions by $2T$ upwards at C.

Now the motion is produced by $2T$ and W, the resistance of the plane

CENTRES OF MASS. 155

being a *passive* force only resists motion. It is obvious then that the

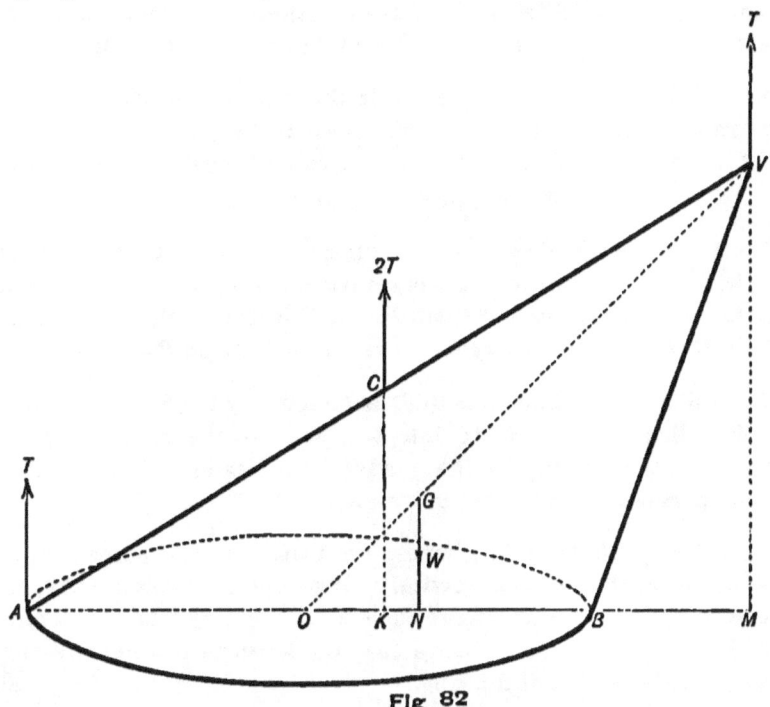

Fig 82

cone will tend to turn about A or B according as C is *right* or *left* of G,

i.e. according as AK is $>$ or $< AN$,

i.e.   ,, $\frac{1}{2}AM$ is $>$ or $< AO + \frac{1}{4}OM$,

i.e.   ,, $\frac{1}{2}AM$ is $>$ or $< AO + \frac{1}{4}(AM - AO)$,

i.e.   ,, AM is $>$ or $< 3AO$.

EXAMPLES.

1. ABC is a triangle, D, E, F are the middle points of its sides, shew that the centre of gravity of the perimeter of ABC coincides with the centre of the circle inscribed in DEF.

2. $ABCD$ is any plane quadrilateral figure, and a, b, c, d are respectively the centres of gravity of the triangles BCD, CDA, DAB, ABC; shew that the quadrilateral $abcd$ is similar to $ABCD$.

3. Prove that the centre of gravity of a wedge, bounded by two similar, equal, and parallel triangular faces and three rectangular faces, coincides with that of six equal particles placed at its angular points.

4. A thin uniform wire is bent into the form of a triangle ABC, and heavy particles of weight P, Q, R are placed at the angular points: prove that if the centre of mass of the particles coincides with that of the wire
$$P : Q : R = b+c : c+a : a+b.$$

5. The perpendiculars from the angles A, B, C meet the sides of a triangle in P, Q, R: prove that the centre of gravity of six particles proportional respectively to $\sin^2 A$, $\sin^2 B$, $\sin^2 C$, $\cos^2 A$, $\cos^2 B$, $\cos^2 C$, placed at A, B, C, P, Q, R, coincides with that of the triangle PQR.

6. A plane quadrilateral $ABCD$ is bisected by the diagonal AC, and the other diagonal divides AC into two parts in the ratio $p : q$; shew that the centre of gravity of the quadrilateral lies in AC and divides it into two parts in the ratio $2p+q : p+2q$.

7. A heavy elliptical ring, whose eccentricity is $\frac{4}{5}$, is suspended with its plane horizontal by three vertical strings, one of which is attached to the end of the minor axis, one to the end of the major axis, and one to the end of a latus rectum. Prove that the tensions respectively are $\frac{1}{4}$, $\frac{1}{4}$, and $\frac{7}{12}$ of the weight of the ring.

8. A triangular table is supported by three legs at the middle points of its sides. A given weight is placed upon it in any position. If weights P, Q, R placed in succession at its angular points will just upset it, prove that $P+Q+R$ is constant.

9. A uniform wire is bent into the form of a circular arc and its two bounding radii, the arc being greater than a semicircle. Shew that if the acute angle between these bounding radii be $\tan^{-1} \frac{4}{3}$, the centre of gravity of the whole wire is at the centre.

10. A triangular lamina is supported at its three angular points and a weight equal to that of the triangle is placed upon it; find the position of the weight if the pressures on the points of support are proportional to $4a+b+c$, $a+4b+c$, $a+b+4c$, where a, b, c are the lengths of the sides of the triangle.

11. Particles are placed at the corners of a tetrahedron respectively proportional to the opposite faces: prove that their centre of gravity is at the centre of the sphere inscribed in the tetrahedron.

12. $ABCD$ is a quadrilateral whose diagonals intersect in O. Parallel forces act at the middle points of AB, BC, CD, DA respectively proportional to the areas AOB, BOC, COD, DOA. Prove that the centre of parallel forces is at the fourth angular point of the parallelogram described on OE, OF as adjacent sides, where E, F are the middle points of the diagonals of the quadrilateral.

13. A solid, consisting of a hemisphere and a right circular cone on opposite sides of the same circular base, is in equilibrium, when placed with any point of the hemisphere on a horizontal plane. If the whole solid can just be included in the sphere of which the hemisphere in question is half, prove that the density of the cone is three times that of the hemisphere.

14. If three uniform rods of the same material but of different thicknesses be formed into a triangle ABC, and if their centre of gravity be at the orthocentre of this triangle, prove that their thicknesses must be proportional to

$$\cos(B-C) - 3\cos A, \quad \cos(C-A) - 3\cos B, \quad \cos(A-B) - 3\cos C.$$

15. The corners of a pyramid are cut off by planes parallel to the opposite sides: if the pieces cut off be of equal weight, prove that the centre of gravity of the remainder will coincide with that of the pyramid.

16. Two uniform heavy rods, AB, BC, rigidly united at B, are hung up by the end A; shew that BC will be horizontal if

$$\sin C = \sqrt{2} \cdot \sin \tfrac{1}{2} B.$$

17. A uniform triangular lamina of weight W is suspended from a fixed point by means of strings attached to its angular points: shew that, unless its plane be vertical, the tensions of the strings are

$$\frac{W \cdot l_1}{\sqrt{\{3(l_1^2 + l_2^2 + l_3^2) - a^2 - b^2 - c^2\}}},$$

and similar expressions; l_1, l_2, l_3 being the lengths of the strings, and a, b, c the sides of the triangle.

18. Find the centre of gravity of a solid sector of a sphere, in which the density at any point varies as the cube of its distance from the centre.

19. A heavy body with a cavity in which lies a small loose heavy sphere, is suspended from a fixed point: shew that, if it hang indifferently in all positions, the form of the cavity must be spherical.

20. Find the centre of mass of the segment of a spheroid cut off by a plane perpendicular to the axis.

21. Shew how to determine the position of the centre of gravity of the area contained between two concentric, similar, and similarly situated ellipses and two straight lines drawn from the common centre.

22. If from a triangle ABC three equal triangles ARQ, BPR, CQP be cut off, the centres of inertia of the triangles ABC, PQR will be coincident.

23. Out of a uniform circular disc, radius a, are cut two circular holes, radii b and c, and centres at distances β, γ from that of the disc, and distance δ from one another. Find where to cut the hole of radius $\sqrt{bc}$, so that the centre of mass of the remainder may be the centre of the disc: if the distance of the centre of this hole from that of the disc be r, shew that

$$r^2 + \delta^2 = (b^2 + c^2)(\beta^2/c^2 + \gamma^2/b^2).$$

24. A rectangular sheet of stiff paper, whose length is to its breadth as $\sqrt{2}$ is to 1, lies on a horizontal table with its longer sides perpendicular to the edge and projecting over it. The corners on the table are then doubled over symmetrically so that the creases pass through the middle point of the side joining the corners and make angles of 45° with it. The paper is then on the point of falling over; shew that it had originally ⅞ of its length on the table.

25. ABC is a triangle; APD, BPE, CPF the perpendiculars from it on opposite sides. Prove that the resultant of six equal parallel forces, acting at the middle points of the sides of the triangle and of lines PA, PB, PC, passes through the centre of the circle which goes through all of these middle points.

26. The inscribed circle of a triangle ABC touches the sides in D, E, F. Prove that the centre of gravity of weights proportional to BC, CA, AB, placed at A, B, C respectively, coincides with the centre of gravity of the same weights placed at D, E, F respectively.

27. Find the centre of gravity of that part of the circumscribing circle of a triangle which lies outside the nine-points circle; and shew that its distance from the centre of the circumscribing circle is half that of the centre of gravity of the triangle.

28. A, B, C, D, E, F are six equal particles at the angles of any plane hexagon, and a, b, c, d, e, f are the centres of gravity respectively of ABC, BCD, CDE, DEF, EFA, and FAB. Shew that the opposite sides and angles of the hexagon $abcdef$ are equal, and that the lines joining opposite angles pass through one point which is the centre of gravity of the particles A, B, C, D, E, F.

29. Find the centre of mass of a solid hemisphere whose density varies inversely as the distance from the centre.

30. A circle whose diameter is equal to the latus rectum of a parabola has double contact with it. Find the position of the centre of mass of the area bounded by the two curves.

31. A triangular lamina ABC hangs at rest from the point A: if $AB = c$, $AC = b$, and S represent the area of the lamina, prove that the tangent of the inclination of BC to the vertical is equal to $4S/(b^2 \sim c^2)$.

32. A smooth solid hemisphere rests with its flat base against a vertical wall and is supported by a string, one end of which is fastened to the vertex of the hemisphere and the other to a point in the wall. Prove that the inclination of the string to the vertical exceeds

$$\tan^{-1}(\tfrac{5 5}{1 8}).$$

33. Find the centre of gravity of the surface of the octant of a sphere.

34. In the side BC of a triangle ABC a point E is taken such that $CE = 4CB \sin^2 18^\circ$; AE is produced to D so that $ED = \tfrac{1}{2}AE \operatorname{cosec} 18^\circ$, and DC is joined. Prove that E is the centre of mass of the figure $ABEDCA$.

35. If the opposite edges of a tetrahedron are equal, prove that the centre of gravity of its six edges and the centre of gravity of its four faces both coincide with the centre of gravity of its volume.

36. If A, B, C be three fixed points, and P any point on a circle whose centre is O, shew that

$$AP^2 \cdot \triangle BOC + BP^2 \cdot \triangle COA + CP^2 \cdot \triangle AOB = \text{constant}.$$

37. From an external point an enveloping cone is drawn to a sphere: prove that the centre of gravity of a uniform solid bounded by the sphere and cone is at a distance $AN^2/4CN$ from the centre of the sphere, where CA is the radius of the sphere from the centre C drawn towards the outer point and cutting the plane of contact in N.

38. The centre of gravity of a solid hexahedron whose faces are triangles is the same as that of five equal weights placed at the corners, and of an equal negative weight placed at the point where the line forming the two trihedral angles cuts the plane of the other three angles.

39. A pack of cards is laid on a table and each projects in direction of the length of the pack beyond the one below it: if each is on the point of tumbling independently of those below it, prove that the distance between the extremities of successive cards will form a harmonical progression.

40. Prove that the sum of the squares of the sides of the triangle, formed by joining the feet of the perpendiculars let fall from a point inside a given triangle on the sides, has its least possible value, when the point is the centre of mass of three particles, at the angles of the given triangle, whose masses are proportional to the squares of the opposite sides.

41. A uniform circular disc of weight nW has a heavy particle of weight W attached to a point on its rim. If the disc be suspended from a point A on its rim, B is the lowest point: and if suspended from B, A is the lowest point. Shew that the angle subtended by AB at the centre is $2\sec^{-1}2(n+1)$.

42. A thin shell is bounded by two similar surfaces; any closed curve being drawn on the surface, prove that the centre of inertia of the included portion of the shell, and the centre of inertia of the solid formed by drawing lines to the boundary from the centre of similitude, are in a line with the centre of similitude and at distances from it which are in the ratio 4 : 3.

43. A frustum is cut from a right cone by a plane bisecting the axis and parallel to the base. Shew that it will rest with its slant side on a horizontal table if the height of the cone bear to the diameter of the base a greater ratio than $\sqrt{7} : \sqrt{17}$.

44. Four weights are placed at four fixed points in space, the sum of two of the weights being given and also the sum of the other two; prove that their centre of mass lies on a fixed plane, and within a certain parallelogram in that plane.

45. A sphere, radius r, rests on three points at equal distances a apart on a horizontal plane. If one of these points be depressed so that the plane containing the three is inclined at an angle θ to the horizon, the sphere will roll off if θ exceed $\sin^{-1}(a/r\sqrt{3})$, but if the point be raised the sphere will roll off if θ exceed $\sin^{-1}\{a/\sqrt{3(4r^2-a^2)}\}$.

46. A hemispherical bowl of radius r rests on a smooth horizontal table, and partly inside it rests a rod of length $2l$, of weight equal that of the bowl. Shew that the position of equilibrium is given by

$$l \sin(a+\beta) = r \sin a = -2r \cos(a+2\beta),$$

where a is the inclination of the base of the hemisphere to the horizon, and 2β is the angle subtended at the centre by the part of the rod within the bowl.

47. Two equal segments are cut from a hollow sphere, and are hung up from a point by two equal strings attached to their rims, so that their convexities are outwards. Prove that, if the lengths of the strings be equal to the diameter of either rim, they are inclined to each other at an angle $= 2\tan^{-1}(\frac{1}{5}\tan\frac{1}{2}a)$, where $2a$ is the angle subtended by either segment at the centre of the sphere.

48. A cone of vertical angle $2a$ is supported by a string passing over two smooth pullies in the same horizontal line, the string being attached to the vertex and to a point in the circumference of the base. Prove that in the position of equilibrium $\sin(a+\theta+\phi) = \frac{3}{2}\cos a \sin \theta \cos \phi$, where θ is the inclination of either portion of the string to the horizon, and ϕ is the angle the base of the cone makes with the vertical.

49. The top of a right cone, semi-vertical angle a, cut off by a plane making an angle β with the axis, is placed on a perfectly rough inclined plane with the major axis of the base along a line of greatest slope of the plane; in this position the cone is on the point of toppling over: prove that the tangent of the inclination of the plane to the horizon has one of the values

$$\frac{4\sin 2a \pm \sin 2\beta}{\cos 2a - \cos 2\beta}.$$

50. A ring is made up of three arcs, BC, CA, AB, of uniform section, but of different metals: uniform rods OA, OB, OC, made of the same metals as BC, CA, AB respectively, but with sectional area double that of the arcs, connect the points A, B, C with the centre O. Find the angles

a, β, γ which BC, CA, AB subtend at O, in order that the centre of gravity of the whole may be at O, and shew that, if ω_1, ω_2, ω_3 be the weights per unit length of BC, CA, and AB respectively,

$(\omega_1 - \omega_2)\tan\tfrac{1}{2}a \tan\tfrac{1}{2}\beta + (\omega_2 - \omega_3)\tan\tfrac{1}{2}\beta \tan\tfrac{1}{2}\gamma + (\omega_3 - \omega_1)\tan\tfrac{1}{2}\gamma \tan\tfrac{1}{2}a = 0.$

51. Find the centre of mass of a spherical surface, over which the density at any point varies as the nth power of the distance from a fixed point on the surface.

52. A solid of uniform density formed by the revolution of a quadrant of a circle about a tangent at one extremity is placed with its vertex and one point in the rim of the base resting on a horizontal plane. Prove that the pressure on the table at the last of these two points is one-eighth of the weight of the right circular cone of the same height, base and density.

CHAPTER V.

FRICTION.

105. WE have hitherto supposed, that the action exerted by one surface in contact with another is necessarily *along* the common normal at the point of contact, in other words that the surfaces are *perfectly smooth*. We have however no experience of bodies except such as do, in certain cases, exert on other bodies forces *inclined* to the common normal at the point of contact, in other words all bodies we are acquainted with are more or less *rough*.

Suppose the following experiment to be made. Take a mass of some material having a plane surface, and fix it so that this surface is horizontal: on it place a portion of some solid material. Now it will be found that, whatever be the materials used, and however highly their surfaces in contact may be polished and lubricated, it is always possible to turn the horizontal surface through a finite angle without the upper body *slipping*, though it may *topple over*.

Let W be the weight of the upper body, α the inclination to the horizon of the plane on which it rests: then

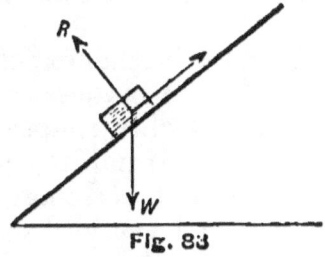

Fig. 83

resolving W into $W\cos\alpha$ perpendicular to the plane and $W\sin\alpha$ along the plane, we infer that as W is counteracted by the action of the plane on the body, this action must consist of two components R ($=W\cos\alpha$) along the common normal, and F ($=W\sin\alpha$) along the plane. The latter force is called the *friction*. We see also that $F/R = \tan\alpha$. When the body is just about to slide the friction exerted is said to be the *limiting* friction.

106. The laws relating to *statical* friction are:

(i) *Friction always acts in the direction opposite to that, in which the point of the surface acted upon, would move, relatively to the other surface, if there were no friction.*

(ii) *The magnitude is always the least possible required for preserving equilibrium, provided this amount does not exceed the limiting friction.*

These laws are *axiomatic* and are particular cases of the general axiom that a *Passive force, being entirely due to the tendency to motion caused by Active forces, only resists such tendency: its direction therefore is always directly opposite to the motion resisted and its magnitude never exceeds the minimum required for preserving equilibrium, and is if possible equal to this minimum.*

107. Let us now make another experiment. As before take a plane surface of some material or other, and on it place blocks of different weights, shapes and sizes, but all made of the same material. If now the plane be gradually inclined in *any* direction more and more to the horizon, it will be found that each and every block, no matter what face it has in contact with the plane, begins to *slide* as soon as a certain inclination of the plane to the horizon is exceeded, but not before; also, that when it does slide, the increase per second in its velocity is constant. This angle though constant for the same pair of materials varies considerably for different pairs. Any block may *topple over* before the others slide.

Let us see what inferences can be drawn from this experiment.

Let α be the inclination of the plane to the horizon, when all the blocks are just about to slide: the friction exerted is in each case limiting, and since $F/R = \tan \alpha$, (Art. 105), the ratio of the limiting friction to the normal pressure is the same for all the blocks. Also since the weights and therefore the normal pressures differ, this ratio is independent of the normal pressure. Since α is the same whatever face of a block rests on the plane, the ratio $F:R$ is independent of the area of the surfaces in contact. Since the increase per second in the velocity of a block is constant, the force on it is constant, i.e. the friction is independent of the velocity.

The above experiment confirms the so-called Laws of *limiting* and *dynamical* friction.

These laws are

(i) *So long as the substances in contact are unaltered, the ratio of the limiting or dynamical friction to the normal pressure is independent of the magnitude of the latter.*

(ii) *So long as the substances in contact are unaltered, the friction is independent of the area of the surfaces in contact.*

(iii) *When motion takes place, the dynamical friction is independent of the relative velocity of the points in contact.*

108. These laws must not be regarded as rigorously true in all circumstances, but only as more or less approximate expressions of the results obtained from the experiments of Coulomb and Morin, who enunciated them. More recent investigations would seem to shew, that in certain circumstances they are very far indeed from expressing the amount of friction exerted.

According to a report, read before the Institution of Mechanical Engineers by Captain Douglas Galton, on experiments made by him on the application of brakes to locomotive-wheels, the friction diminishes as the velocity increases beyond a certain limit, and is also less after it has been exerted for some time than when first applied. In the experiments of

Morin and Galton the surfaces in contact were not lubricated in any way. Before the same Institution in 1883, Mr Beauchamp Tower read a report on some experiments made by himself on a thoroughly lubricated journal revolving in bearings. These experiments shewed that in certain circumstances the friction per square inch was nearly independent of the normal pressure and that it increased with the velocity of revolution. A rise in temperature was accompanied by a reduction in the friction, though this might be caused by the lubricant becoming more efficient. Professor Thurston states that from his own experiments, he inferred that the friction at first diminished as the velocity increased and then increased again.

As however we are only concerned with statical friction we may take laws (i) and (ii) as giving fairly accurately the friction in the cases which we shall have to consider.

109. *Def.* The ratio of the limiting friction to the normal pressure, which ratio we see by laws (i) and (ii) is constant, is called the *coefficient of friction* for the pair of materials in contact. The angle the total action makes with the common normal at the point of contact is termed the *angle of friction*, provided the limiting friction is exerted.

Hence (Art. 105) the coefficient of friction is equal to the tangent of the angle of friction.

The *coefficient of dynamical friction* is the ratio of the friction to the normal pressure when motion is actually taking place. It is found by experiment to be less than the corresponding coefficient of statical friction, in other words, there is more resistance when motion is just about to take place than when it is actually taking place.

Ex. 1. If the smallest force which will move a given block weighing 3 lbs. along a given horizontal plane be $\sqrt{3}$ lbs.; find the greatest angle at which the plane may be inclined to the horizon without the block sliding. *Ans.* $30°$.

Ex. 2. If a weight of 14 lbs., when placed on a rough plane inclined at an angle of $60°$ to the horizon, slides down, unless a force of at least 7 lbs. acts on it up the plane, what is the coefficient of friction? *Ans.* ·73.

Ex. 3. If a weight of 4 lbs. is just on the point of slipping down a rough plane, inclined at an angle of $45°$ to the horizon, when a force of 2 lbs. acts up the plane, find the least force which will move the weight up the plane, when the inclination is $30°$ to the horizon. *Ans.* 3·01 lbs.

Ex. 4. Weights of 4 and 5 lbs. respectively, connected by a light rigid rod, are placed on a rough inclined plane, with the rod parallel to a line of greatest slope. If the coefficient of friction between the 4 lb. weight and the plane be ·6 and that between the other weight and the plane ·42, find the greatest inclination of the plane to the horizon, consistent with equilibrium. *Ans.* $\tan^{-1} \cdot 5$.

Ex. 5. Find the greatest angle at which a plane may be inclined to the horizon so that three equal weights whose coefficients of friction are ·5, ·6, ·7, respectively, may when connected by strings rest on it without sliding. The weights are supposed placed along a line of greatest slope so that each is rougher than the one next below it. *Ans.* $\tan^{-1} \cdot 6$.

Ex. 6. A uniform ladder rests in limiting equilibrium, with its lower end in contact with a rough horizontal plane and its upper end with a smooth vertical wall. If λ be the angle of friction and a the angle the ladder makes with the vertical, prove that $\tan a = 2 \tan \lambda$.

Ex. 7. If everything is as in Ex. 6, except that the wall is as rough as the ground, prove that $a = 2\lambda$.

Ex. 8. Two equal heavy rings, P and Q, slide on two rough rods inclined at the same angle a to the horizon: a string connecting P and Q passes through an equal ring. Shew that if P and Q are each on the point of slipping down, the inclination (θ) of either part of the string to the vertical is given by the equation

$$\tan \theta = 3 \tan (a - \lambda),$$

where λ is the angle of friction.

Ex. 9. A body is resting on a rough inclined plane of inclination a, the angle of friction being ϕ which is greater than a. Shew that the ratio of the least force which will drag the body up the plane to the least force which will drag it down is $\sin (\phi + a) : \sin (\phi - a)$.

Ex. 10. One end of a uniform rod is on a rough inclined plane to which the rod is perpendicular: at the other end is applied a force parallel to the plane: if the rod be in equilibrium, prove that the coefficient of friction cannot be less than half the tangent of the plane's inclination.

Ex. 11. Two equal rough balls lie in contact on a rough horizontal table: another equal ball is placed on them so that the centres are in a vertical plane: find the least angle of friction (1) between the upper and lower balls and (2) between the lower balls and the table, in order that they may be in equilibrium. *Ans.* (1) $15°$, (2) $\tan^{-1}\frac{1}{3}(2-\sqrt{3})$.

Ex. 12. A rectangular block of cast iron whose base is 2 feet square and weight 10 tons, rests on a floor of cast iron (coefficient of friction ·16). A rope is attached to it at such a height above the floor, and pulled with such a force in a direction making an angle measured upwards of $30°$ with the floor, that the block is on the point of sliding and tumbling: find (1) the height of the point to which the rope is attached and (2) the tension of the rope.

Ans. (1) 6·83 ft., (2) 1·68 tons.

Ex. 13. Two weights resting on a rough inclined plane, whose inclination a is greater than the angle of friction λ, are connected by a string which passes over a smooth peg on the plane: shew that the least possible ratio of the less to the greater is $\sin(a-\lambda)/\sin(a+\lambda)$.

Ex. 14. Two equal heavy rings hang on a rough horizontal rod, and are connected by a string of length c which supports an equal ring: find the greatest possible distance between the first two rings.

Ans. $3\mu c/(1+9\mu^2)^{\frac{1}{2}}$.

110. *Def.* Let a cone be described, having its vertex at the point of contact of two surfaces, the common normal for axis, and the angle of friction as semi-vertical angle. This cone is called the *Cone of Friction*.

The Laws of statical friction, given in Arts. 106, 107, are all included in the following statement. *If all the other forces, external and internal, acting on the point of contact be compounded into a single resultant R, the action of the surface in contact will be equal and opposite to R, whatever be the latter's magnitude or direction, provided its line of action does not lie without the cone of friction.*

Hence a body in contact with rough surfaces will be in equilibrium, provided that to ensure its being so, it is not necessary to assume that the total action at any point of contact lies outside the corresponding cone of friction.

FRICTION. 169

It should be noticed that the cone of friction is always drawn so that its concavity is *towards* the body, the action on which we are considering.

111.* *To find the relation between the tensions at the ends of a light string stretched over a rough surface, and on the point of slipping.*

Let $APQRZ$ be the string, which is on the point of slipping from Z to A. Let the points $A, B,...P, Q,$

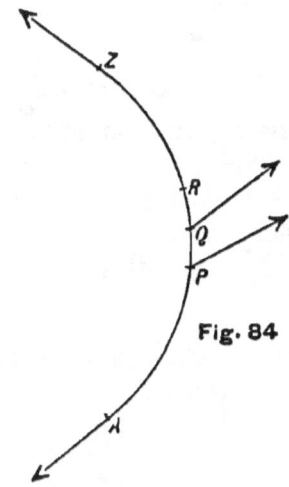

Fig. 84

$R,...Y, Z$ be taken so that the ultimately indefinitely small angles between the tangents at consecutive points are each equal to θ. Let λ be the angle of friction, μ the coefficient of friction.

Let us consider a small portion, PQ, of the string. It is kept in equilibrium by the tensions at P and Q and the resultant action of the surface.

As in Art. 81, construct a force-diagram $Oab...pqr...yz$, such that $Oa, Ob,...Op, Oq, Or,...Oy, Oz$ represent the tensions at $A, B,...P, Q, R,...Y, Z$ respectively. Join $ab, bc,...pq, qr,...yz$. These last will represent the resultant actions of the surface on $AB, BC,...PQ, QR,...ZY$ respectively.

As the portion PQ is on the point of slipping from Q to P, the resultant action on it makes with the normal

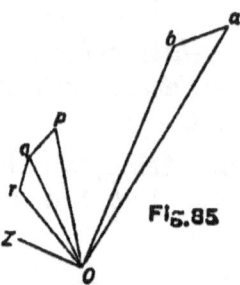

Fig. 85

at either P or Q an angle differing from λ by an indefinitely small quantity, and on that side of the normal by which it will most assist the tension at Q. Hence in each of the triangles Oab, Opq, &c. the angles at O are equal, and the angles Oab, Opq, &c. are each equal to $\frac{1}{2}\pi - \lambda$ ultimately; the triangles therefore are all similar to one another.

Let n be the infinitely large number of portions AB, BC, &c. of the string: then $n\theta = $ a finite angle, a say.

Then
$$Oa = Ob\,\frac{\cos(\lambda - \theta)}{\cos \lambda},$$

$$\ldots = \ldots\ldots\ldots\ldots$$

$$Op = Oq\,\frac{\cos(\lambda - \theta)}{\cos \lambda},$$

$$\ldots = \ldots\ldots\ldots\ldots$$

$$Oy = Oz\,\frac{\cos(\lambda - \theta)}{\cos \lambda}.$$

$$\therefore Oa = Oz\left(\frac{\cos(\lambda - \theta)}{\cos \lambda}\right)^n = Oz\,(\cos \theta + \sin \theta \tan \lambda)^n;$$

$$\therefore \log Oa = \log Oz + \frac{n}{2}\log(1 - \sin^2 \theta) + n\log(1 + \mu \tan \theta)$$

$$= \log Oz - \tfrac{1}{2}\left(n\sin^2\theta + \frac{n}{2}\sin^4\theta + \&c.\right)$$
$$+ \mu n \tan\theta - \frac{\mu^2 n \tan^2\theta}{2} + \&c.$$

But $\quad n\sin^2\theta = \alpha \cdot \dfrac{\sin^2\theta}{\theta} = 0$ ultimately,

$$n\sin^4\theta = \alpha \cdot \frac{\sin^4\theta}{\theta} = 0 \quad \ldots\ldots\ldots\ldots$$

$$n\tan\theta = \alpha \cdot \frac{\tan\theta}{\theta} = \alpha \quad \ldots\ldots\ldots\ldots$$

$$n\tan^2\theta = \alpha\tan\theta = 0 \quad \ldots\ldots\ldots\ldots$$

and each series is convergent,

$\therefore\ \log Oa = \log Oz + \mu\alpha$ ultimately;

$\therefore\ Oa = Oz \cdot \epsilon^{\alpha\mu}$;

$\therefore$ tension at $A = \epsilon^{\alpha\mu} \cdot$ tension at Z.

If the string be in one plane, the curve $abcd...z$ will be a plane one, and as the tangent at every point makes a constant angle with the line joining the point with O, the curve is an equiangular spiral. The ratio of Oa to Oz might therefore be obtained from the known properties of that curve.

If the string be not in one plane, the curve $abc...$ will not be a plane one; it can however be made so, without altering the distance of any point from O, by turning each of the triangles Oab, Obc, &c. about a side terminating in O, until they are all in one plane, when the curve becomes an equiangular spiral, and the ratio of Oa to Oz can be obtained as suggested above.

Ex. A string, attached to a weight of 10 lbs. which rests on a rough horizontal plane, passes over a rough peg, and just supports a weight of 5 lbs. at the other end: whereas if the string be coiled once round the peg, a weight of 80 lbs. can be supported by it without the other weight slipping. Find the coefficient of friction of the plane, assuming that the position of the string between the 10 lb. weight and the peg is horizontal. *Ans.* ·25.

112.* We shall sometimes be required to solve problems of the following kind. A system of bodies is in equilibrium under certain conditions: a gradual change occurs in one or more of these conditions—e.g. the coefficient of friction at one or more points of contact of the bodies is gradually diminished, some external force is gradually altered in magnitude or direction, or the position of one of the bodies is gradually altered. When this gradual change reaches a certain stage equilibrium is no longer possible, and it is required to ascertain the way in which equilibrium is broken, in other words, the nature of the initial motion of the different bodies. The actual way in which equilibrium is broken must satisfy the following conditions. The various forces acting on the different bodies, when such a motion is about to take place, must be able to adapt themselves so as to satisfy the necessary conditions of equilibrium, without in any way violating the laws relating to passive forces: they must also be incapable of satisfying the necessary conditions of equilibrium, if the change in the initial conditions increase still further.

We shall generally proceed by considering the different ways in which it is conceivable equilibrium might be broken, without violating the geometrical conditions. If only one of these satisfies the above conditions, it is the way required; if more than one satisfy it, it is beyond the limits of this treatise to obtain a solution of the problem.

The following rule will often enable us to solve such problems. *If it is inconceivable that equilibrium can be broken, except by one of the bodies either turning about or sliding past a point of contact with another body, the former motion will actually take place, provided it does not involve the assumption that the total action at the point in question lies outside the cone of friction.*

FRICTION. 173

This rule is a deduction from the axiomatic law relating to passive forces (Art. 106). For if we suppose the body connected with the other body at the point of contact by a smooth joint, it can only *turn about* that point. If now motion be on the point of taking place, the first body will be about to turn about the joint, which will exert some action on it. If this action does not necessarily lie outside the cone of friction, it could be exerted at the point of contact if no joint existed, i.e. the motion is the same without the joint as with. On the other hand, if the action at the joint be outside the cone of friction, it could not be exerted unless the joint existed, i.e. equilibrium is about to be broken by the body *sliding past* the point of contact in question. When it is necessary to assume that the action at the joint is exerted *along a generator* of the friction-cone, the question cannot be solved by the above rule, as it shews that slipping is about to occur at the point at the same instant as rolling.

Exs. 5, 6, 7, 9, 10 are illustrative of this principle and should be studied attentively.

ILLUSTRATIVE EXAMPLES.

Ex. 1. A uniform rod MN rests with its ends in two fixed straight grooves OA, OB, in the same vertical plane, and making angles a, β with

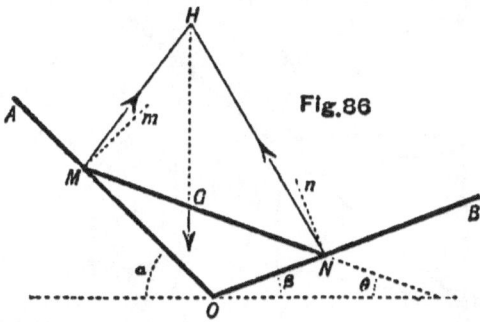

Fig. 86

the horizon: prove that, when the end M is on the point of slipping down AO, the tangent of the inclination of MN to the horizon is

$$\frac{\sin(a - \beta - 2\epsilon)}{2\sin(\beta + \epsilon)\sin(a - \epsilon)}.$$

174   STATICS.

Let θ be the inclination of MN to the horizon, when M is on the point of slipping down AO.

Draw Mm, Nn, normals to OA, OB, respectively. Since the point M is on the point of moving down AO, the limiting friction is exerted at M in the direction MA, and the direction of the total action of OA on the rod makes the angle ϵ with Mm, on the side towards A.

Similarly, because N is on the point of slipping *up* OB, the total action of OB on the rod makes the angle ϵ with Nn on the side towards O.

Let the lines of action of the forces on MN at M and N meet in H; then, (Art. 61), H is vertically above G, the middle point of the rod.

Join HG.

$$MG : GH = \sin MHG : \sin HMG = \sin (\alpha - \epsilon) : \sin (\tfrac{1}{2}\pi + \theta - \alpha + \epsilon);$$

also $\quad NG : GH = \sin NHG : \sin GNH = \sin (\beta + \epsilon) : \sin (\tfrac{1}{2}\pi - \theta - \beta + \epsilon);$

$$\therefore \sin (\alpha - \epsilon) : \cos (\alpha - \epsilon - \theta) = \sin (\beta + \epsilon) : \cos (\beta + \epsilon + \theta).$$

Hence we obtain $\quad \tan \theta = \dfrac{\sin (\alpha - \beta - 2\epsilon)}{2 \sin (\alpha - \epsilon) \sin (\beta + \epsilon)}.$

Ex. 2. A glass rod is balanced partly in and partly out of a cylindrical tumbler with the lower end resting against the vertical side of the tumbler.

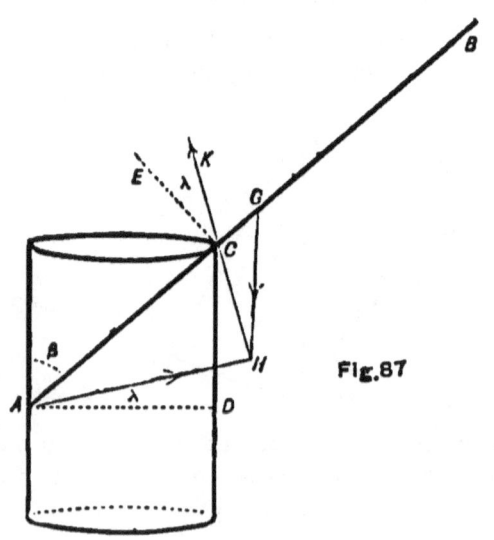

Fig. 87

FRICTION.

If a and β are the greatest and least angles which the rod can make with the vertical, prove that the angle of friction, λ, is

$$\tfrac{1}{2} \tan^{-1} \frac{\sin^3 a - \sin^3 \beta}{\sin^2 a \cos a + \sin^2 \beta \cos \beta}.$$

Let AB be the rod, G its centre of mass. Let C be the point of the edge of the tumbler on which AB rests. Draw AD normal to the tumbler at A and CE perpendicular to the rod at C.

(i) When AB makes the *smallest* possible angle with the vertical, and is therefore on the point of slipping *into* the tumbler.

Since A is on the point of slipping down, the action there on AB is in the direction AH, which makes the angle λ with AD on the side towards C. Similarly, the action at C on the rod is in the direction CK, which makes the angle λ with CE, on the side away from A.

Let KC and AH meet in H, which must therefore be vertically below G.

Join GH. Let a be the diameter of the tumbler, and let $AG = c$.

$$AG : AH = \sin AHG : \sin AGH = \cos \lambda : \sin \beta,$$

and $\qquad AH : AC = \sin ACH : \sin AHC = \cos \lambda : \sin (2\lambda + \beta);$

$$\therefore AG : AC = \cos^2 \lambda : \sin \beta \sin (\beta + 2\lambda),$$

$$\therefore c : a \operatorname{cosec} \beta = \cos^2 \lambda : \sin \beta \sin (\beta + 2\lambda).$$

(ii) When AB makes the *greatest* possible angle with the vertical and is therefore on the point of slipping *out of* the tumbler.

By reasoning as before, we should have AH and CK on the sides of AD and CE respectively, opposite to those they were on in the first case, and we should arrive at the result obtained there, except that for β we must write a, and for λ, $-\lambda$.

The result would therefore be

$$c : a \operatorname{cosec} a = \cos^2 \lambda : \sin a \sin (a - 2\lambda);$$

∴ eliminating c and a, we have

$$\sin^2 \beta \sin (\beta + 2\lambda) = \sin^2 a \sin (a - 2\lambda);$$

$$\therefore \tan 2\lambda = \frac{\sin^3 a - \sin^3 \beta}{\sin^2 a \cos a + \sin^2 \beta \cos \beta}.$$

Ex. 3. A uniform rectangular board $ABCD$ rests with the corner A against a rough vertical wall and its side BC on a smooth peg, the plane of the board being vertical and perpendicular to that of the wall. Show that, without disturbing the equilibrium, the peg may be moved through a space $\mu \cos a\,(a \cos a + b \sin a)$ along the side with which it is in contact, provided μ do not exceed a certain value: a being the angle BC makes with the wall, and a, b the lengths of AB, BC respectively.

Let G be the intersection of diagonals, i.e. the centre of mass of the board. Let P be a position of the peg when there is equilibrium.

The forces acting on the board are, its weight vertically downwards

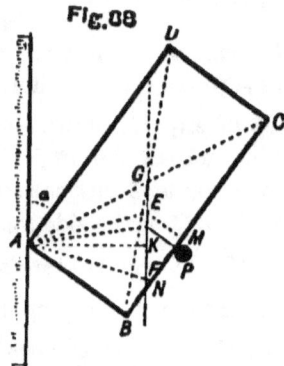

Fig. 68

through G, the reaction of the peg through P and at right angles to BC, and the reaction of the wall through A.

The necessary and sufficient condition of equilibrium is that these three forces should meet in a point, as the magnitudes of the reactions at P and A will adapt themselves to secure equilibrium, if the above condition holds.

Let the first two forces meet in K; join AK, which is therefore the direction of the reaction of the wall. But AK is not a possible direction of the reaction at A, if it makes with the normal to the wall an angle greater than $\tan^{-1} \mu$.

Draw AE and AF making with the normal on either side of it the angle $\tan^{-1} \mu$, and meeting GK in E and F. Draw EM, FN perpendicular to BC. The condition of equilibrium is then that K should lie between E and F, i.e. that P should lie between M and N. We may therefore, without disturbing the equilibrium of the board, move the peg through the space MN along BC.

And $MN = EF \cos a = 2\mu \cos a \times$ horizontal distance of G from wall

$$= \mu \cos a\,(a \cos a + b \sin a).$$

We have assumed above that M and N are both between B and C: if either lies beyond B or C, as the peg cannot be moved off the board without disturbing the equilibrium, it can only be moved along that part of MN which lies between B and C. It is obvious that if μ be greater than a certain value, either M or N will not lie between B and C.

Ex. 4. If one cord of a sash window breaks, find the coefficient of friction of the sash in order that the other weight may still support the window.

Let $ABCD$ be the window, W its weight acting at G its centre of mass.

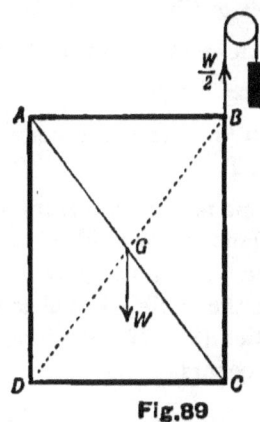

Fig. 89

We assume that the window fits *loosely* in the sash, so that there will be contact at only *one* point on each side; these will be A and C respectively.

The unbroken cord at B supplies a force $\tfrac{1}{2}W$ vertically upwards; the resultant of this and W is $\tfrac{1}{2}W$ vertically downwards at A. Hence in order that equilibrium may be possible, the action at C must be along CA, i.e. the coefficient of friction at C must be not less than $\tan ACD$.

Ex. 5.* A right circular cone, vertical angle $2a$, rests with its base on a rough horizontal plane: a string is attached to the vertex and pulled in

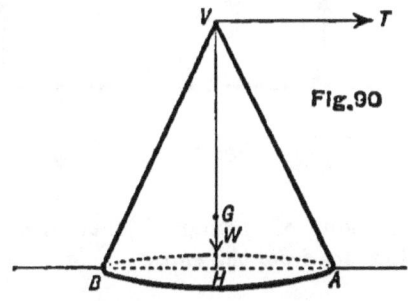

Fig. 90

a horizontal direction with a gradually increasing force: determine how the equilibrium will be broken.

Let VAB be a vertical section of the cone containing the direction of the string. Let T be the tension of the string when equilibrium is about to be broken, and W the weight of the cone.

The different ways in which it is conceivable equilibrium may be about to be broken are

(1) the cone being *lifted bodily* from the plane,

(2) the cone *tilting*, with one point of the base resting on the plane,

(3) the cone *sliding* along the plane.

(1) is impossible, as in that case the cone would be in equilibrium under the action of W and T.

If (2) take place, the cone is in equilibrium under the action of W, T, and the reaction of the plane at the point of contact, which must therefore be A; also the action at A must pass through V, i.e. along AV. This is only possible when the angle AV makes with the vertical, i.e. a, is less than the angle of friction (λ). If, therefore, a be $<\lambda$, (2) takes place, (Art. 112): if a be $>\lambda$, (3) occurs.

Ex. 6.* A heavy straight rod, whose sectional area varies as the distance from one end, rests on a rough horizontal plane. At the other end, perpendicularly to its length and in the horizontal plane, a force is applied of gradually increasing magnitude: prove that the rod begins to turn about its middle point.

Let AB be the rod, l its length, B the end at which the force is applied. Let S be the force at B, when motion is just about to take place.

Let us investigate whether or no there is a point in AB about which

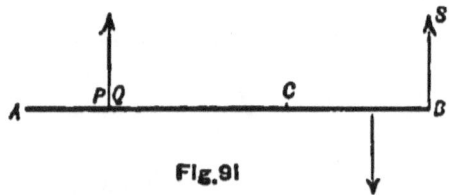

Fig. 91

the rod may be on the point of turning. If there is such a point, let C be it, and let x be its distance from the end A.

The friction on any point of the rod between C and B acts in the opposite direction to S, and that on any point in CA in the same direction as S.

The friction on any small portion PQ is $\mu \times$ weight of PQ, and therefore the total friction on AC is $\mu \times$ weight of AC, and acts at the c. m. of AC. Also the total friction on CB is $\mu \times$ weight of CB, and acts at the c. m. of CB.

By Art. 102, the weight of $AC = \kappa x^2$, and that of AB is κl^2: also the distance from A of the c. m. of AC is $\frac{2}{3}x$, that of the c. m. of AB is $\frac{2}{3}l$. The weight of the remainder CB is therefore $\kappa (l^2 - x^2)$, and the distance of its c.m. from A is $\frac{2}{3}(l^3 - x^3)/(l^2 - x^2)$.

∴ taking moments about A for the equilibrium of the rod, we have

$$\mu\kappa(l^2 - x^2) \cdot \frac{2}{3}\frac{(l^3 - x^3)}{(l^2 - x^2)} - \mu\kappa x^2 \cdot \tfrac{2}{3}x = Sl \ldots\ldots\ldots\ldots\ldots(1).$$

Also resolving at right angles to the rod, we have

$$\mu\kappa(l^2 - x^2) - \mu\kappa x^2 = S \ldots\ldots\ldots\ldots\ldots\ldots(2);$$

∴ eliminating S from equations (1) and (2)

$$2(l^3 - 2x^3) = 3l(l^2 - 2x^2);$$

∴ $4x^3 - 6lx^2 + l^3 = 0$.

Since $x = \frac{1}{2}l$ satisfies this equation, the rod must begin to turn about its middle point, as it will turn about it rather than slip there (Art. 112).

Ex. 7.* A square lamina is supported in a horizontal position by means of four rough pegs on which its angles A, B, C and D rest. A horizontal force is applied at C at right angles to AC and gradually increased until

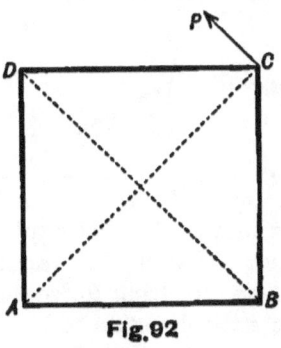

Fig. 92

it moves the lamina. Shew that, if the pressures on the pegs be equal, the lamina will begin to turn about the angle A.

We know (Art. 112) that the square will be on the point of turning about A, provided that all the necessary equations of equilibrium can be satisfied on such an assumption, without requiring the friction exerted at A to be the maximum.

Let P be the applied force which will cause the lamina to be just on the point of motion. Let O be the point of intersection of the diagonals of the square. Let Q be the maximum friction that can be exerted at any of the corners of the square—then if rotation is about to take place about A, the force at D will be Q along DC, that at C, Q opposite to P, and that at B, Q along CB.

Taking moments about A, we have
$$P \cdot AC = Q(AD + AC + AB),$$
$$\therefore P = Q(1 + \sqrt{2}).$$

The friction at A must be equal to the resultant of the other four forces, its magnitude is therefore

$$\sqrt{\{(P - Q - Q\sqrt{2})^2 + (\tfrac{1}{2}Q\sqrt{2} - \tfrac{1}{2}Q\sqrt{2})^2\}}, \text{ i.e. zero.}$$

A is the point therefore about which the square will begin to turn.

Ex. 8. A heavy particle is placed on a rough inclined plane whose inclination is equal to the angle of friction: a thread is attached to the particle and passed through a hole in the plane which is lower than the particle, but not in the line of greatest slope: shew that if the thread be very slowly drawn through the hole the particle will describe a straight line and a semi-circle in succession.

Let O be the hole; OA the horizontal line in the inclined plane

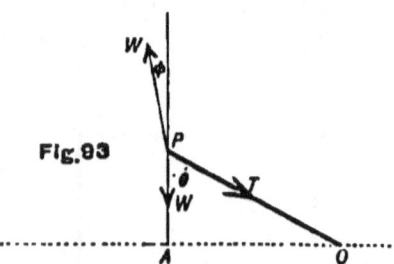

Fig. 93

through O. Let P be the particle, W the resolved part of its weight in the plane. The maximum friction that can be exerted on it is W therefore. Let θ be the angle the string PO makes with a line of greatest slope. Let ϕ be the angle the direction of motion at any instant, and therefore the friction, makes with a line of greatest slope. Let T be the tension of the string.

FRICTION.

(1) When P is above OA.

The resolved parts of forces down the line of greatest slope
$$= W + T\cos\theta - W\cos\phi.$$
Those perpendicular to the same line $= T\sin\theta - W\sin\phi.$

Since the particle is drawn *very slowly*, each of these forces must be indefinitely small. Therefore ϕ and T are both indefinitely small. Hence the particle moves down a line of greatest slope, until it reaches A.

(2) When P is at A.

The forces now are $W - W\cos\phi$ and $T - W\sin\phi$, whence we infer that $\phi = 0$, i.e. the particle moves off initially at right angles to OA. The particle, however, cannot remain any longer in the same line of greatest slope, and since it must always be approaching O, it describes a curve, which has a line of greatest slope as tangent at A, and which passes through O.

(3) When P is below OA.

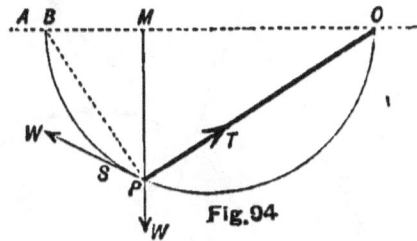

Fig. 94

In this case we deduce, as before, that
$$W(1 - \cos\phi) - T\cos\theta = 0,$$
$$W\sin\phi - T\sin\theta = 0.$$

The solution $T = 0 = \phi$ is inadmissible here, since we know that the particle cannot continue to move down a line of greatest slope.

Eliminating T, we have $\dfrac{1 - \cos\phi}{\sin\phi} = \cot\theta,$

$$\therefore \tan\tfrac{1}{2}\phi = \cot\theta,$$
$$\therefore \tfrac{1}{2}\phi = \tfrac{1}{2}\pi - \theta.$$

Draw PB perpendicular to OP, meeting OA in B: describe a semicircle through P, on OB as diameter.

Let SP be the tangent at P to this circle. Then
$$\angle SPM = \angle SPB + \angle BPM = \pi - 2MPO = \phi.$$

182   STATICS.

Therefore the direction of motion at P is along the tangent to the circle, i.e. the next point to P in P's path is on the circle. Similarly the next consecutive point to that and so on.

Hence the semi-circle is the particle's path, and as this is true always so long as P is below OA, the semi-circle must pass through A, i.e. A and B are coincident.

Ex. 9.* A uniform heavy beam AB is placed with the end A upon a rough horizontal plane and a point C of its length touching a rough heavy sphere whose point of contact with the plane is D. Prove that if there is equilibrium the magnitude of the friction at each of the three points A, C, D will be the same. If the coefficient of friction be the same at each point, the point at which slipping is most likely to take place will be A or C, according as A and D lie on the same or opposite sides of the vertical through B.

Let O be the centre of the sphere, G the middle point of the beam.

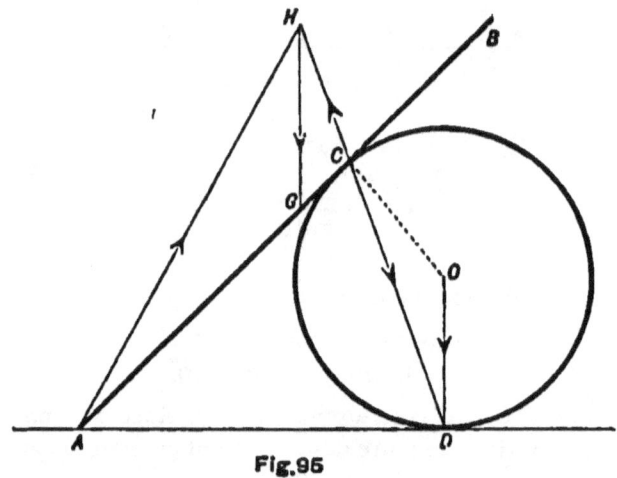

Fig. 95

Considering the equilibrium of the sphere and beam together, since the horizontal forces acting on them are the frictions at A and D respectively, they must be equal.

Also from the equilibrium of the sphere, by taking moments about O, we deduce that the friction at C = that at D = that at A.

Let us suppose that the sphere is slowly moved away from A, until equilibrium is about to be broken; what will be the nature of the motion

FRICTION. 183

which is about to happen? Of the three forces acting on the sphere, two, the weight and the reaction at D act through D, therefore the third, the action at C, is along CD, and the action at D is within the angle CDO. Hence slipping cannot be about to occur at D, as then the angle CDO, and therefore the angle OCD, would be greater than the angle of friction, which is impossible, as it is the angle the action at C makes with the normal.

Let DC produced meet the vertical through G in H: join AH. AH is the direction of the action on AB at A.

Hence either the angle AH makes with the vertical, or OCD must be the angle of friction, as slipping must occur at either A or C.

The slipping occurs at A or C

    according as $\angle AHG$ is $>$ or $<$ $\angle OCD$, i.e. $\angle ODC$, i.e. $\angle GHC$,

    according as G is nearer A or D,

    according as A and B are on the same or opposite sides of the vertical through D.

Ex. 10.* A block in the shape of a rectangular parallelepiped of weight W rests with one edge horizontal on a rough inclined plane;

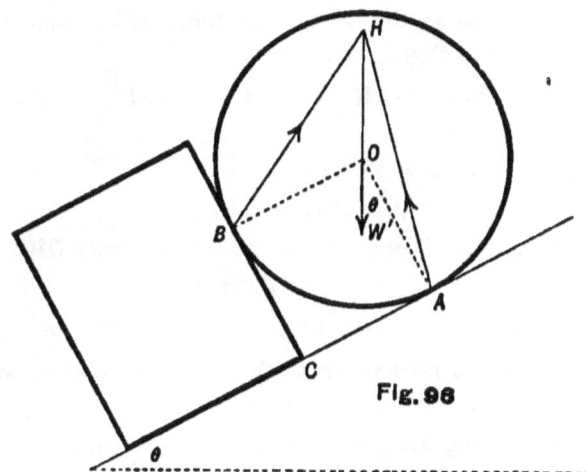

Fig. 96

against the block rests a rough sphere (W') whose radius is less than the thickness of the block. The inclination of the plane is gradually increased until equilibrium is no longer possible: shew that if the block tilt, the sphere will slide or roll along the plane according as the limiting inclination (θ) of the plane to the horizon is $>$ or $< \frac{1}{4}\pi$; and shew also that if the block slide, the sphere will slide or roll according as λ (the angle of friction) is $>$ or $< \frac{1}{4}\pi$, and that in the last case, θ is given by the equation $W \sin(\lambda - \theta) = W' \sin\theta (\cos\lambda - \sin\lambda)$, λ being supposed the same everywhere.

Let O be the centre of the sphere, A and B the points where it touches the plane and block respectively: C the point of the block nearest to A. Let the inclination of the plane (θ) be such that equilibrium is just on the point of being broken.

There are only two motions of the block conceivable,

(1) turning about its lowest edge,

(2) sliding down the plane.

Whichever of these two ways the block moves, the sphere will either

(α) slip at B and roll at A, or

(β) roll at B and slip at A.

The actions at A and B on the sphere must meet in the vertical through O, in the point H say: join AH, BH, these will be the directions of the respective reactions.

As we have seen either (α) or (β) must occur, one of the angles OBH, OAH must equal λ, and as the other angle must be less than λ, it is the greater angle of the two that is equal to λ.

We shall prove that $\angle OBH$ is $>$ or $< \angle OAH$, according as θ is $<$ or $> \frac{1}{4}\pi$.

If θ is $< \frac{1}{4}\pi$, $\angle AOH$ is $> \angle BOH$,

$$\therefore AH \text{ is } > BH,$$

and as $BO = OA$, and OH is common to the two triangles OBH, OHA,

$$\angle OBH \text{ is } > \angle OAH.$$

Similarly it can be shewn that if θ is $> \frac{1}{4}\pi$, $\angle OAH$ is $> \angle OBH$.

Hence whether (1) or (2) happen to the block the sphere will roll or slide at A, according as θ is $<$ or $> \frac{1}{4}\pi$.

If (2) and (β) happen, $\theta = \lambda$, and λ is therefore $> \frac{1}{4}\pi$.

To find θ when (2) and (α) happen,

Let R be the normal reaction at B, then taking moments about A for the equilibrium of the sphere,

$$W' a \sin\theta = R(1 + \tan\lambda) a.$$

FRICTION.

Resolving along the plane for the equilibrium of the block

$$W \sin \theta + R = (W \cos \theta + R \tan \lambda) \tan \lambda.$$

$\therefore W (\sin \theta - \cos \theta \tan \lambda)(1 + \tan \lambda) = W' \sin \theta (\tan^2 \lambda - 1),$

$\therefore W \sin (\lambda - \theta) = W' \sin \theta (\cos \lambda - \sin \lambda).$

Since θ cannot be greater than λ, this equation shews that if λ be $> \frac{1}{4}\pi$, (2) and (a) cannot happen.

EXAMPLES.

1. A body is supported on a rough inclined plane by a force acting along it. If the least magnitude of the force, when the plane is inclined at an angle a to the horizon, be equal to the greatest magnitude when the plane is inclined at an angle β, shew that the angle of friction is $\frac{1}{2}(a - \beta)$.

2. Two equal particles on two inclined planes are connected by a string which lies wholly in a vertical plane perpendicular to the line of junction of the planes, and passes over a smooth peg vertically above this line of junction. If, when the particles are on the point of motion, the portions of the string make equal angles with the vertical, shew that the difference between the inclinations of the planes must be twice the angle of friction.

3. A uniform rod is resting on a rough inclined plane, and is moveable on the plane about one end which is fixed: shew that when it is about to slip it makes with the line of greatest slope the angle $\sin^{-1}(\mu \cot a)$.

4. Spheres whose weights are W, W' rest on different and differently inclined planes. The highest points of the spheres are connected by a horizontal string perpendicular to the common horizontal edge of the two planes above it. If μ, μ' the coefficients of friction are such that each sphere is on the point of slipping down, $\mu W = \mu' W'$.

5. Two equal particles rest upon two equally rough inclined planes, being connected by a string passing over a smooth pulley at the common vertex, the vertical plane which contains the string being at right angles to each inclined plane. If the weight of one particle be increased by a certain amount the system is on the point of motion, and if instead the weight of the other particle be decreased by the same amount the system is again on the point of motion in the same direction as before. Prove that the difference of the inclinations of the two planes is double the angle of friction.

6. A lamina is suspended by three strings from a point: if the lamina be rough, and the coefficient of friction between it and a particle placed upon it be constant, shew that the boundary of possible positions of equilibrium of the particle on the lamina is a circle.

7. A uniform heavy rod is in equilibrium in a rough spherical cup; and the length of the rod subtends a right angle at the centre of the sphere; find the greatest angle the rod can make with the horizon in terms of the angle of friction.

8. Two fixed pegs are in a line inclined at a given angle a to the horizon. A rough thin rod rests on the higher and passes under the lower, the higher peg being lower than the centre of gravity of the rod. The distance of that point from the pegs being a and b respectively, shew that when the rod is on the point of motion $(b+a)\mu = (b-a)\tan a$.

9. Prove that the direction of the least force required to draw a carriage is inclined at an angle θ to the ground, where $a\sin\theta = b\sin\phi$, a being the radius of the wheels, b of the axles, and $\tan\phi$ the coefficient of friction of the axles.

10. A light string is placed over a rough vertical circle, and a uniform heavy rod, whose length is equal to the diameter of the circle, has one end attached to each end of the string, and rests in a horizontal position. Find within what points on the rod a given mass may be placed, without disturbing the equilibrium of the system: and shew that the given mass may be placed anywhere on the rod, provided the ratio of its weight to that of the rod does not exceed $\frac{1}{2}(e^{\mu\pi}-1)$, where μ is the coefficient of friction between the string and the circle.

11. Two particles of unequal mass are tied by fine inextensible strings to a third particle. They lie on a rough horizontal plane with the strings stretched at a given angle to each other. Find the magnitude and direction of the least horizontal force which, applied to the third particle, will move all three.

12. An equilateral triangle, of uniform material, rests with one end of its base on a rough horizontal plane and the other against a smooth vertical wall: shew that the least angle its base can make with the horizontal plane is given by the equation $\cot\theta = 2\mu + \frac{1}{3}\sqrt{3}$, μ being the coefficient of friction.

13. Two weights P, Q are connected by a string and rest one on each face of a double inclined plane, the string passing over the common vertex, which is smooth: at first P is about to slip downwards and when

the weights are interchanged, it is found that P is still just about to slip downwards: shew that if λ, λ' are the angles of friction for the two planes and a, β the angles they respectively make with the horizon, then
$$\cos a \cos \lambda' = \cos \beta \cos \lambda.$$

14. Two rough spheres, the larger of which is fixed, rest on a rough horizontal plane, and a uniform board rests symmetrically upon the top of them, its centre of gravity being midway between the points of contact: shew that, if $\tan \lambda'$ and $\tan \lambda$ be the coefficients of friction between the board and the larger and smaller spheres respectively, and motion be about to take place at both points of contact, $\tan (\lambda' - \lambda) = \sin 2\lambda$.

15. Two rings, each of weight w, slide upon a vertical semi-circular wire, diameter horizontal and convexity upwards. They are connected by a light string of length $2l$ (supposed less than the diameter $2a$) on which is slipped a ring of weight W. Shew that when the two rings are as far apart as possible, the angle $2a$ subtended by them at the centre is given by $(W+2w)^2 \tan^2 (a+\epsilon) (l^2 - a^2 \sin^2 a) = W^2 a^2 \sin^2 a$, ϵ being the angle of friction.

16. An isosceles triangular prism is placed with its edge horizontal and its base on a rough inclined plane, the inclination of which is gradually increased: shew that the prism will tumble or slide according as μ is $>$ or $< 3c/2h$. c is the base of a section perpendicular to the edge and h the height.

17. Two hemispheres, of radii a and b, have their bases fixed to a horizontal plane, and a plank rests symmetrically upon them. If μ be the coefficient of friction between the plank and either hemisphere, the other being smooth, prove that, when the plank is on the point of slipping, the distance of its centre from its point of contact with the smooth hemisphere is equal to $(a \sim b)/\mu$.

18. A disc in the shape of a sector of a circle lies on a rough table (μ) and is fastened at the centre by a peg. Shew that the least force applied along any tangent to the sector necessary to turn it round is to the weight of the disc as $2\mu : 3$.

19. A rod rests partly within and partly without a box in the shape of a rectangular parallelepiped, and presses with one end against the rough vertical side of the box and rests in contact with the opposite smooth edge. The weight of the box being four times that of the rod, shew that if the rod be about to slip and the box about to tumble at the

same instant, the angle the rod makes with the vertical is

$$\tfrac{1}{2}\lambda + \tfrac{1}{2}\cos^{-1}(\tfrac{1}{2}\cos\lambda),$$

where λ is the angle of friction.

20. Three equal heavy rough cylinders are placed in contact along generating lines, lying on a horizontal plane: and two other such cylinders are similarly placed upon them: find the frictions and reactions at the instant when the system is bordering on motion.

21. A sphere (radius a) whose centre of gravity is distant c from its centre, rests in limiting equilibrium on a rough plane, which is inclined at an angle a to the horizon: shew that the sphere may be turned through the angle $2\cos^{-1}(a/c \csc a)$ and still be in limiting equilibrium.

22. Assuming that the limiting friction consists of two parts, one proportional to the pressure, and the other to the surface in contact, shew that if the least force which can support a rectangular parallelepiped, whose edges are a, b, and c on a given inclined plane be P, Q, R, when the faces in contact are bc, ca, ab respectively, then

$$(Q-R)bc + (R-P)ac + (P-Q)ab = 0.$$

23. A rough rod rests over a rough sphere, one end of the rod pressing on a rough horizontal plane, on which the sphere rests. Shew that there will be limiting equilibrium for the whole system when the rod makes an angle $2\lambda_2$ with the plane, if the weight of the sphere is to the weight of the rod in the ratio $\sin(\lambda_2 - \lambda_1) : \sin(\lambda_2 + \lambda_1)$, where λ_1 is the angle of friction between the rod or sphere and the plane, and λ_2 the angle of friction between the rod and sphere.

24. A rectangular lamina rests in a vertical plane with the middle point of one side in contact with a rough peg, coefficient of friction 2, and a point in the opposite side in contact with a smooth peg. If the line joining the pegs make an angle a with the vertical, and the sides in contact with the pegs an angle θ, when the lamina is just about to slip, shew that $\tan(\theta + a) = 1 - 2\tan\theta$.

25. A uniform heavy rod PQ is in equilibrium with its ends on a rough parabola whose axis is vertical and vertex downwards: shew that the line joining the intersection of the tangents to the parabola at P, Q to the intersection of the normals makes with the vertical an angle not greater than the angle of friction.

26. A pair of equal rods AB, AC are hinged together at A and have rings at B, C: these rings are free to slide along fine rough straight wires OB', OC' in the same vertical plane equally inclined at an angle a to the vertical. Shew that if λ be the angle of friction, in the limiting positions of equilibrium the angle between the rods is either

$$2\tan^{-1} 2\cot(a-\lambda) \text{ or } 2\tan^{-1} 2\cot(a+\lambda).$$

27. A right-angled isosceles triangular lamina rests with its base angles on the arc of a rough circular wire whose plane is vertical and radius equal to either of the equal sides of the triangle. If the equal sides be horizontal and vertical in the limiting position of equilibrium the coefficient of friction is $\frac{1}{4}\{\sqrt{17}-3\}$.

28. Two uniform rods of equal weight, but different lengths, are jointed together and placed in a vertical plane over two rough pegs in the same horizontal line: if a, β be the inclinations of the rods to the horizon, θ that of the reaction at the hinge, prove that when the rods are on the point of slipping, $2\tan\theta = \cot(\beta+\lambda) - \cot(a-\lambda)$, where λ is the angle of friction.

29. An ellipse is placed with its plane vertical and major axis horizontal so that one of its vertices A rests against a rough vertical wall. P is a point on the wall vertically above A, and a string of length $2l$ which has its extremities fastened at the foci S, H passes through P. Find the least value of the coefficient of friction consistent with the equilibrium of the ellipse.

30. A uniform ladder (length $2a$) rests at an angle a to the vertical against a smooth horizontal rail at a height h from the ground. If λ be the angle of friction, between the ground and the ladder, shew that a man of weight n times that of the ladder may ascend a distance along the ladder, $\{2(n+1)h\sin\lambda \sec(a-\lambda)\operatorname{cosec} 2a - a\}/n$, without the ladder slipping.

31. A uniform beam AB lies horizontally upon two others at points A and C; prove that the least horizontal force applied at B, in a direction perpendicular to BA which is able to move the beam, is the less of the two forces $\mu W \cdot \dfrac{b-a}{2a-b}$ and $\dfrac{\mu W}{2}$, where $AB=2a$, $AC=b$, $W=$ weight of beam, and $\mu=$ coefficient of friction.

32. A uniform rod of mass M rests in a horizontal position with its ends on the circumference of a rough vertical circle and subtends an angle

$2a$ at the centre. An insect of mass m starts from the middle point of the rod and crawls gently towards one end. Prove that if the angle ϵ of friction be less than $45°$ it will be able to reach the end of the rod without disturbing the equilibrium provided $\sin 2\epsilon > m \sin 2a/(M+m)$.

Examine the case when $\epsilon > 45°$.

33. A number of equally rough particles are knotted at intervals on a string, one end of which is fixed to a point on an inclined plane. Shew that, all the portions of the string being tight, the lowest particle is in its highest possible position when they are all in a straight line making an angle $\sin^{-1}(\tan \lambda \cot a)$ with the line of greatest slope, λ being the angle of friction, and a the inclination of the plane to the horizon. Shew also that, if any portion of the string make this angle with the line of greatest slope, all the portions must do so too.

34. A uniform rod, length $2a \sin a$, is placed within a rough vertical circle, radius a, and is on the point of motion, the coefficients of friction at its upper and lower ends are $\tan \lambda'$, $\tan \lambda$: prove that if θ be the inclination to the vertical of the line joining the centre of the circle to the centre of the rod
$$\tan \theta = \frac{\sin (\lambda + \lambda')}{2 \cos (\lambda + a) \cos (\lambda' - a)}.$$

Examine the case when $a + \lambda = \tfrac{1}{2}\pi$.

35. One end of a heavy rod AB can slide along a rough horizontal rod AC to which it is attached by a ring; B and C are joined by a string: if ABC be a right angle when the rod is just on the point of slipping, and a the angle between AB and the vertical, shew that the coefficient of friction is
$$\frac{\sin a \cos a}{1 + \cos^2 a}.$$

36. A circular lamina, whose centre of gravity is at an excentric point, rests in a vertical plane supported by the loop of a rough string which is attached to two fixed points. If the lamina be on the point of slipping and the radius containing its centre of gravity be inclined at right angles to the radius bisecting the portion of the string in contact with the circle, the angle of contact ϕ, is given by
$$\frac{1 + \epsilon^{\mu\phi}}{1 - \epsilon^{\mu\phi}} \sin \frac{\phi}{2} = \frac{a}{c},$$
a being the radius of the circle and c the distance of its centre from its centre of gravity.

FRICTION. 191

37. A rough circular disc of radius a has its centre of gravity at a distance b from the centre, and rests in a vertical plane on two pegs placed at a distance apart $< 2\sqrt{(a^2-b^2)}$ and $> 2b$ in a horizontal line: shew that equilibrium is possible for all positions of the centre of gravity provided the angle of friction be not less than $\sin^{-1}(b/a)$.

38. Two equal heavy rods AB, BC, each of a length $2a$, joined together at B, hang with AB resting on a rough peg P. If μ be the coefficient of friction, and 2α the angle between the rods, shew that AB will slip on the peg if $PB < a \cos\alpha (\cos\alpha - \mu \sin\alpha)$ or $> a \cos\alpha (\cos\alpha + \mu \sin\alpha)$.

39. A uniform isosceles triangular lamina rests in a limiting position of equilibrium in a vertical plane between two rough pegs in the same horizontal line: prove that $3c \cos\lambda \sin(2\theta + 2\alpha - \lambda) = 2p \sin 2\alpha \cos(\theta + \alpha)$, where θ is the inclination of one side to the horizon, λ the angle of friction, 2α the vertical angle of the triangle, p the perpendicular from the vertex on the base, and c the distance between the pegs.

40. Three rough particles of masses m_1, m_2, m_3 are rigidly connected by light smooth wires meeting in a point O, such that the particles are at the vertices of an equilateral triangle whose centre is O. The system is placed on an inclined plane of slope α, to which it is attached by a pivot through O; prove that it will rest in any position if the coefficient of friction for none of the particles be less than

$$\frac{\tan\alpha}{m_1+m_2+m_3}(m_1^2+m_2^2+m_3^2-m_2m_3-m_3m_1-m_1m_2)^{\frac{1}{2}}.$$

41. A cylindrical rod with hemispherical ends rests in a vertical plane against two equally rough planes, one horizontal, the other vertical: determine the limiting position of equilibrium, and shew that if the coefficient of friction be not less than the ratio of the length of the straight part of the rod to the total length, it will rest in any position.

42. A uniform heavy rod of given length rests perpendicularly and horizontally across two rough parallel horizontal rails which support the rod at a quarter of its length from each end. One end of the rod is pulled perpendicularly by a string in a downward direction making an angle θ with the vertical: shew that the rod will move at both points of support at the same time when $\theta = \tan^{-1} 2\mu$; and in this case find the tension of the string.

43. To the ends of a heavy rod are attached rings which slide on the circumference of a rough vertical circle. Find the force perpendicular to its direction, acting at a given point of it which will just move the rod when in any position: and prove that for all positions it will be greatest when the rod is inclined to the horizon at an angle

$$\tan^{-1}(\cot 2\epsilon + \cos 2a \operatorname{cosec} 2\epsilon),$$

where $2a$ is the angle subtended by the rod at the centre and $\tan \epsilon$ the coefficient of friction.

44. A straight uniform rod of length $2c$ is placed in a horizontal position as high as possible within a hollow rough sphere of radius a. Prove that the line joining the middle point of the rod to the centre of the sphere makes with the vertical an angle $\tan^{-1} \mu a / \sqrt{(a^2 - c^2)}$.

45. A semi-circular arch, composed of an odd number of equal and similar smooth blocks, is constructed upon a rough horizontal plane: prove that the number of blocks must be 3: and that the coefficient of friction must be not $< \frac{1}{3}\sqrt{3}$. Also prove that the ratio of the internal to the external arch must not be $>$ the positive root of the equation

$$2\sqrt{3}\,(x^2 + x + 1) + \pi\,(2x^2 - x - 3) = 0.$$

If the blocks, except the key stone, be rough, and if their number be n, greater than 3, prove that the angle of friction at the pth joint from the base must be not $< \cot^{-1}\{(n - 2p)\tan \pi/2n\} - p\pi/n$.

46. Two particles of equal weight w connected by a rod without weight rest on a rough plane inclined to the horizontal at an angle a: the coefficient of friction $\rho' \tan a$ for one particle is less, and that for the other $\rho \tan a$ greater, than $\tan a$. Prove that, when both are on the point of moving, if in the plane a triangle ABP be constructed whose sides AB, BP, PA are 2, ρ', ρ, and O be the middle point of AB which is drawn in a line of greatest slope, then OP is the direction and $OP \cdot w \sin a$ is the tension of the rod.

47. An elliptical cylinder placed in contact with a vertical wall and a horizontal plane is just on the point of motion when its major axis is inclined at an angle a to the horizon. Determine the relation between the coefficients of friction of the wall and plane: and shew from your result that if the wall be smooth, and a be equal to $45°$, the coefficient of friction between the plane and cylinder will be equal to $\frac{1}{2}e^2$, where e is the eccentricity of the transverse section of the cylinder.

48. Two equal spheres rest on a rough horizontal plane, the distance between their centres being c: and a third sphere rests on them: prove that the normal pressure between either sphere and the upper one is equal to half the weight of the upper sphere, and that the necessary and sufficient condition of equilibrium is $a+b > \tfrac{1}{2}c \operatorname{cosec} 2\epsilon$, where ϵ is the angle of friction, and a, b the radii of the spheres.

If this condition is not fulfilled how will the lower spheres begin to move?

49. An elliptic lamina of eccentricity e rests upon a perfectly rough equal and similar lamina, the two bodies being symmetrically situated with respect to their common tangent at the point of contact. If a be the inclination of the major axis of the fixed ellipse to the horizon, and θ be the inclination, measured in the same direction, of the major axis of the moving ellipse in a position of equilibrium, then

$$\sin \tfrac{1}{2}(\theta+a) = e^2 \sin \theta \cos \tfrac{1}{2}(\theta-a).$$

50. A chain is formed by $2n$ rods, equal in length and weight, smoothly jointed together. The two extremities can move by rings on a rough horizontal rod, coefficient μ. Shew that in the limiting position of equilibrium the inclination of either of the upper rods to the vertical is $\tan^{-1} \dfrac{2n\mu}{2n-1}$.

51. A rough elliptic cylinder rests with its axis horizontal upon the ground and against a vertical wall, the ground and the wall being equally rough; shew that the cylinder will be on the point of slipping when its major axis plane is inclined at an angle of $\tfrac{1}{4}\pi$ to the vertical if the eccentricity of its principal section be $\sqrt{\{2\sin\lambda\,(\sin\lambda+\cos\lambda)\}}$, where λ is the angle of friction.

52. An elliptic lamina moveable about its focus in a vertical plane rests against a smooth inclined plane, the major axis of the ellipse being horizontal. The lower surface of the plane is rough and rests just on the point of moving on a horizontal table. If a, b be the semi-axes of the ellipse, and p the perpendicular from the centre on the inclined plane, shew that the coefficient of friction is $\sqrt{\{(p^2-b^2)/(a^2-p^2)\}}$.

53. A circular ring of weight W hangs in a vertical plane over a rough peg, and to the lowest point of the ring a string is fastened. It is kept always horizontal in the plane of the ring, and its tension is gradually increased from zero. Prove that the ring will slip on the peg

when the tension of the string reaches the value $W \tan \frac{1}{2}\{\sin^{-1}(3\sin\epsilon) - \epsilon\}$, ϵ being the angle of friction; and explain what happens if $3\sin\epsilon > 1$.

If the tension be still further increased to a given value T, find the position of equilibrium.

54. A ring of diameter a is fixed with its plane making an angle α with the vertical, and a uniform rough cylinder is supported by being slipped through the ring: prove that the length of the cylinder must be not less than

$$a\cos\theta \cdot \frac{\cos(\theta \pm a - \lambda)}{\sin(\theta \pm a)\sin\lambda},$$

where λ is the angle of friction, and θ is the inclination to the axis of the cylinder of a plane section whose major axis is equal to a. (The sign to be taken in the above expression depends on whether the cylinder and ring make angles with the vertical on the same or opposite sides.)

55. A cylinder is laid on a rough horizontal plane, and is in contact with a rough vertical wall, the coefficients of friction being equal; a string, coiled round it at right angles to the axis, passes over a fixed pulley and sustains a weight which is gradually increased until equilibrium is broken. Determine the nature of the initial motion.

56. Two uniform beams, of the same material and thickness but of different lengths, rest each with one end on a rough horizontal plane, and their other ends connected by a smooth joint. If equilibrium be about to be broken shew in what way it will happen.

57. Two weights, P, Q, whose coefficients of friction are μ_1, μ_2, each less than $\tan\alpha$, on a rough inclined plane of angle α, are connected by a string which passes through a fixed pulley A in the plane. Prove that if the angle PAQ be the greatest possible the squares of the weights of P, Q, are to one another as $1 - \mu_2^2\cot^2\alpha$ is to $1 - \mu_1^2\cot^2\alpha$.

58. A rough rod is laid on a horizontal table and is acted on by a horizontal force perpendicular to its length. Find about what point the rod will begin to turn, the point of application of the force trisecting the rod.

59. A cubical uniform block is placed on a rough inclined plane and has two of its faces vertical: it is attached by a string parallel to a line of greatest slope of the plane passing from the middle point of its upper horizontal edge to the middle point of the nearest horizontal edge of

another equal similarly situated cube. If μ (less than unity) be the coefficient of friction for the lower block, the equilibrium will be broken when the inclination of the plane to the horizon is given by $2\mu = 3\tan\theta - 1$, by the higher cube tumbling over, provided the friction coefficient for the higher block be great enough.

60. A heavy rod, of length $2l$, rests horizontally on the inside rough surface of a hollow circular cone, the axis of which is vertical and the vertex downwards. If $2a$ is the vertical angle of the cone, and if the coefficient of friction is less than $\cot a$, prove that the greatest height of the rod, when in equilibrium, above the vertex of the cone is

$$l \cot a \cdot \left\{ \frac{1 + \cos^2 a + \sin^2 a \sqrt{(\sin^2 a + 4\mu^2)}}{2(1 - \mu^2 \tan^2 a)} \right\}^{\frac{1}{2}}.$$

61. A cubical block, and a cylinder whose diameter is equal to a side of the cube, are laid upon a rough plane, and are attached to each other by a cord coiled round the middle of the cylinder, and fixed to the middle point of one of the edges of the cube which is parallel to the axis of the cylinder. If the plane be then slowly raised (the cube being uppermost) until equilibrium is broken, what will be the nature of the initial motion?

62. Two particles A and B of weight W are connected by a thin weightless rod and placed on a rough inclined plane at an inclination to the line of greatest slope, the coefficient of friction for each particle being μ. A force F is applied to A the lower particle in the direction BA and its direction gradually turned through an angle θ in the plane. Find the nature of the initial motion of the system. If the particles be placed along a line of greatest slope, prove that both will slip when

$$\cos\theta = \frac{F^2 + 4W^2 \sin a (\sin a - \mu \cos a)}{2FW(\mu \cos a - 2\sin a)},$$

and find the limits between which F must lie when $a < \tan^{-1}\tfrac{1}{2}\mu$.

63. Two hemispheres (centres A, B and weights W_1 and W_2) are placed with their rims on a rough horizontal table and in contact, and a rough sphere (centre C and weight W) rests on them, of such a radius that ACB is a right angle. The system is on the point of moving: shew that the sphere will begin to slip over the larger hemisphere, whilst the larger or the smaller hemisphere will begin to slip according as

$$(W_1 - W_2)\sin\epsilon < \text{ or } > W\sqrt{2}\cos(a + \tfrac{1}{4}\pi)\sin(a + \epsilon),$$

where $\epsilon - \tfrac{1}{4}\pi$ is the angle of friction, and a is the angle CAB.

64. A cylindrical rod with hemispherical ends and another cylinder are in contact on a rough plane, the axis of the former is vertical, that of the latter horizontal. The radius of the horizontal cylinder is such that the other touches it at a point in the rim of its upper hemispherical end. The horizontal cylinder is gradually moved along the plane in a direction perpendicular to its axis, the two remaining in contact: shew that equilibrium is no longer possible unless λ be $> \frac{1}{4}\pi$, and if λ be $> \frac{1}{4}\pi$, equilibrium is impossible when the rod makes an angle $> \theta$ with the vertical, where θ is given by the equation $h \cos \theta = h + (h+a) \cos 2\lambda$.

Explain how the equilibrium is broken in this case.

$a =$ the radius of the hemispherical ends, $2h =$ the length of the generating lines of the rod, and $\lambda =$ the angle of friction, supposed the same everywhere.

CHAPTER VI.

VIRTUAL WORK.

113. *Def.* If the point A at which a force P is acting be displaced to any point B, the distance AB is called the *displacement* of the point.

If from B, BN be drawn perpendicular to P's line of action, the product $P \cdot AN$ is called the *Work done* by

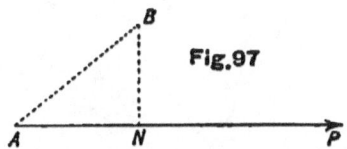

Fig. 97

the force P during the displacement. If N falls on that side of A towards which P acts, the work is said to be positive, if on the other side, negative. We may say then that the product of the force into the projection of the displacement, along the direction of the force, gives the *algebraical* as well as the *numerical* value of the work done during the displacement.

When the work done is *negative*, it is often said to be done *against* the force.

If the displacement does not really take place but is only *imagined* to do so, it is said to be a *Virtual Displacement*, and the work which would be done during such a displacement is called the *Virtual Work*.

STATICS.

114. Prop. *If a particle acted on by any system of forces receive any virtual displacement whatever, the algebraical sum of the virtual work done by the different forces during the displacement is equal to the virtual work done by the resultant.*

Let O represent the actual position of the particle, O' the position to which it is supposed displaced; let P_1, P_2,

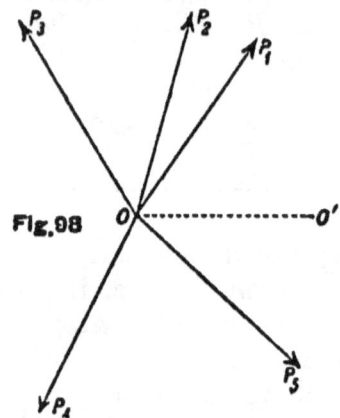

Fig. 98

P_3, &c., be the forces acting on the particle; θ_1, θ_2, θ_3, &c., the angles their directions make with OO'.

The A. S. of the virtual work done by P_1, P_2, P_3, &c.

$= P_1 \cdot OO' \cos \theta_1 + P_2 \cdot OO' \cos \theta_2 + P_3 \cdot OO' \cos \theta_3 + \ldots$

$= OO' \cdot (P_1 \cos \theta_1 + P_2 \cos \theta_2 + \ldots)$

$= OO' \times$ A. S. of the resolved parts of the forces in direction OO'

$= OO' \times$ resolved part of the resultant in direction OO'

$=$ projection of OO' along the direction of the resultant $\times$ the resultant

$=$ the virtual work done by the resultant.

It should be observed that in the above proposition the displacement the particle receives is virtual, and entirely unrestricted both as regards magnitude and direction.

VIRTUAL WORK.   199

Cor. *If a particle in equilibrium under the action of any system of forces receive any virtual displacement whatever, the algebraical sum of the virtual work done by the different forces is zero.*

115. If a system of particles be in equilibrium under the action of external and internal forces, and any number of particles of the system receive *any virtual displacements whatever*, we have seen that the A. S. of the virtual work done by the forces on each particle is zero: hence the A. S. of the virtual work done by *all the forces, external and internal*, is zero.

Prop. *If a system of particles in equilibrium under the action of any system of external forces together with internal forces, receive any indefinitely small virtual displacement whatever, which does not alter the configuration formed by the particles, the* A. S. *of the virtual work done by the external forces alone is zero, or more strictly speaking, is of an order higher than that of the displacement.*

In this proposition the displacements which the particles receive are very much restricted as compared with that in the theorem of Art. 114: here the displacement must be indefinitely small, there it might be of any magnitude: the displacements, too, of the different particles are also so connected, that if the particles formed a rigid body, these displacements would not involve any alteration in its shape or size, but only an alteration of its position as a whole.

We shall first prove that if the displacements be of this character, the virtual work done by any internal force

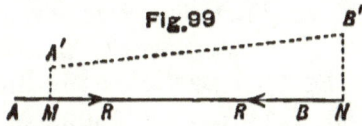

(the action exerted by B) on the particle A is equal and opposite in sign to that done by the reaction exerted by A

on the particle B. Let R be the action exerted by B on the particle A, along AB in the direction indicated, then the reaction exerted by A on B is in the opposite direction. Let A', B' be the points to which A, B are supposed displaced, then by the conditions relating to the nature of the displacements, the angle (θ) between $A'B'$ and AB is small, and the length $A'B' = AB$.

Draw $A'M$, $B'N$ perpendicular to AB.

Then $MN = A'B' \cos \theta = A'B'$ ultimately, $= AB$,
$$\therefore AM = BN.$$

The virtual work done by action R on $A = R \cdot AM$: that done by reaction R on $B = -R \cdot BN = -R \cdot AM$.

Hence the A. S. of the virtual work done by any action and the corresponding reaction is zero. But the internal forces consist entirely of pairs, each pair being made up of an action and the corresponding reaction: therefore the A.S. of the virtual work done by *all* the internal forces is zero, since that of each pair is so. We have seen too that the A. S. of the virtual work done by *all the forces*, both external and internal, is zero, so that that of the virtual work done by the *external forces alone* must be zero also.

In obtaining this result we have neglected quantities depending on the powers higher than the first of the displacement, so that strictly speaking the A. S. of the virtual work done by the external forces is not zero, but of an order higher than the first power of the displacement.

Cor. If in any system of forces in equilibrium there are two forces equal to one another, and acting in opposite directions along the straight line joining the particles on which they respectively act, the two forces will not enter into the equation of virtual work, provided the virtual displacements of the two particles produce no alteration in the length of the line joining them, or at any rate one of the second order only. Hence, if we have two bodies in contact, and the virtual displacement does not alter the points in contact, the action and reaction between the two bodies will not appear in the equation for the two bodies together. Also, if two particles are connected by an inextensible string or rod, and they receive displacements which do not involve breaking or bending the string or rod, the tension of the string, or in the case of the rod, the tension or thrust, whichever it exerts, will not enter into the equation of virtual work for both particles together. This may easily be extended to the case of two particles connected by an inextensible string which passes round a smooth fixed body: for the distance between them measured along the string is constant, so long as the string neither slackens nor breaks.

116. In applying the above proposition to the case of a rigid body, we may suppose the displacement any slight displacement of the body as a whole not involving any change of shape or size. If we wish to ascertain the internal forces between one portion of a body and another, we may suppose that the first portion is displaced as a whole, without any displacement of the remainder, in which case the actions of this last portion on the first will enter into the equation of virtual work.

In solving problems by the principle of virtual work, it is often convenient to make such a displacement that a force, whose magnitude we do not wish to ascertain, may not enter into the equation of virtual work. In that case the virtual work done by that particular force must be zero, or at any rate of a higher order in small quantities than the displacement. For that to be the case the particle on which the force acts must be virtually displaced in a direction making with the force, either a right angle or an angle differing from a right angle by an indefinitely small quantity.

From the way in which the principle of virtual work is sometimes stated, the student is apt to get the idea that it is only for certain kinds of displacements, and for displacements of a certain extent that the principle holds. It is quite true that in order to avoid introducing certain forces it is often convenient to make only certain displacements such as those mentioned in Arts. 115 and 117. It must however be borne in mind that, by the first paragraph of Art. 115, the principle in its most general form may be applied to *any body whatever*, under the action of *any system* of forces.

In the following propositions, it is understood that the displacements are indefinitely small.

117. Prop. *The work done by the tension of an inextensible string or rod, when one end is fixed and the other attached to a particle which is displaced so that the string or rod is neither broken nor bent, is ultimately of an order higher than the first.*

It is obvious that the particle can only move in a direction which is ultimately at right angles to the rod or string, i.e. at right angles to the tension.

Prop. *If a body resting in contact with any smooth curve or surface, receive a displacement by sliding along the curve or surface, the work done by the reaction of the curve or surface on the body is ultimately of an order higher than the first.*

In this case the particle situate at the point of the body touching the curve or surface is the one on which the reaction acts, and this point moves along a tangent to the surface or curve, i.e. at right angles to the normal along which the reaction acts.

Prop. *If a body resting in contact with any surface, not necessarily smooth, receive a displacement by rolling along the surface, the work done by the reaction of the surface is ultimately of an order higher than the first.*

Let A be the common point of the body and the

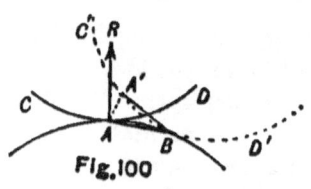

Fig. 100

surface: let the body be rolled so that the point originally at A comes to A', and B becomes the point of contact.

Then the arcs AB, $A'B$ are small, and therefore the corresponding chords; also the angle between these chords is small, so that the base AA', which is the displacement of A, is of a higher order than AB.

Cor. If the body partly roll and partly slide along a smooth surface, it is clear that the displacement is compounded of the two displacements of rolling and sliding, and is therefore of an order higher than the first, since each of the latter is so.

We have already seen that if two bodies in contact receive such virtual displacements that their points of contact remain the same, the action and reaction between the two do not appear in the equation of virtual work for the two bodies: neither will they if in addition to these displacements one rolls along the other, or if the bodies be smooth, one slides or partly slides and partly rolls along the other. For either set of displacements alone will not bring these forces into the equation, therefore a combination will not do so.

118. As an illustration of the application of the principle of virtual work, we will by means of it prove the theorem of Art. 81, viz., that the tensions at the ends of a weightless string stretched over a smooth surface are equal.

Let A, B be the points where the string leaves the surface, and let T, T' be the tensions at the ends P, Q respectively.

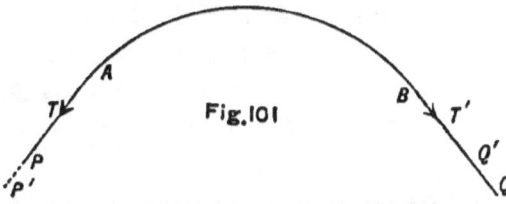

Fig. 101

We shall not interfere with the equilibrium of the string if we suppose it to lie in a groove cut in the surface, so that when pulled at one end, it must move along the groove. Let the virtual displacement which is given to the string be produced by pulling the end P in the direction AP to P', so that the end Q must move along QB to a point Q' such that $QQ' = PP'$. As each portion of the string in contact with the surface moves at right angles to the action of the surface on it, no work is done by the actions of the surface on the string, and the algebraical sum of the virtual work done is $T \cdot PP' - T' \cdot QQ'$, which is therefore zero; i.e. $T = T'$, since $PP' = QQ'$.

Apply the principle of virtual work to the solution of the following examples:

Ex. 1. The algebraical sum of the moments about any point in their plane of a number of coplanar forces in equilibrium is zero.

Ex. 2. Two small rings of equal weight slide on a smooth wire in the shape of a parabola, whose axis is vertical and vertex upwards; they will be in equilibrium if connected by an inextensible string which passes over a smooth peg placed at the focus.

Ex. 3. Two equal uniform rods freely jointed at their ends rest, one on each of two smooth pegs which are in a horizontal line. Shew that the inclination (θ) of either rod to the vertical is given by the equation

$$a \sin^3 \theta = c,$$

where a is the length of each rod, and c the distance between the pegs.

Ex. 4. Shew that in Ex. 27, page 92, the weight of each beam is proportional to the tangent of the angle, which the line joining the centre of the semicircle with the corresponding point of contact of the beam makes with the horizontal.

Ex. 5. Two equal uniform rods of the same material and thickness have two ends connected by a smooth hinge, and their other ends are attached to small rings which slide on a smooth horizontal wire. Find the position of equilibrium when a circular disc whose weight equals that of either rod, is placed between them so that each rod touches its circumference.

Ans. Each rod makes with the vertical the angle (θ) given by the equation $2a \sin^3 \theta = r \cos \theta$, where a is the length of a rod and r the radius of the disc.

Ex. 6. A tripod formed of three equal uniform rods, three ends being connected by a common joint, and the other three connected, each with the other two, by equal strings, rests with the joint uppermost on a smooth horizontal plane. Shew that the tension of each string is $Wc/3h$, W being the weight of a rod, c the length of each string, and h the height of the joint above the plane.

Ex. 7. Prove that the total work done against gravity in raising or lowering any number of bodies is equal to the work done in raising a weight equal to the total weight through the height moved through by the centre of gravity.

Ex. 8. A heavy elastic string rests in the shape of a necklace round a smooth right circular cone whose axis is vertical: shew that its radius is $a + wa^2/\lambda \tan a$, where λ is the modulus of elasticity, $2\pi a$ the length of the string when unstretched, w its weight per unit length, and a the semi-vertical angle of the cone.

Ex. 9. Two small rings of equal weight attracting with a force varying as the distance, slide on a smooth parabolic sloped wire, whose axis is vertical and vertex upwards: shew that if they are in equilibrium in *any* symmetrical position, they are so in *every* one.

Ex. 10. A frame is formed of four uniform rods, freely jointed at their ends and forming a parallelogram $ABCD$. The frame is hung over a peg at A and the points B, D are connected by a weightless rod. Shew that the thrust of the rod : the weight of the frame $= BD : 2AC$.

119. We shall now prove the *converse* of the principle of virtual work for a single particle, i.e.

If the algebraical sum of the virtual work done by a system of forces acting on a particle be zero for every displacement whatever, the particle is in equilibrium.

For let (fig. 98) the forces be P_1, P_2 &c., O the particle, OO' any virtual displacement, θ_1, θ_2, &c. the angles P_1, P_2, &c. make with OO'. Then, since the algebraical sum of the virtual work done by the forces $= 0$,

$$P_1 . OO' \cos \theta_1 + P_2 . OO' \cos \theta_2 + \ldots = 0,$$
$$\therefore OO' (P_1 \cos \theta_1 + P_2 \cos \theta_2 + \ldots) = 0,$$
$$\therefore P_1 \cos \theta_1 + P_2 \cos \theta_2 + \ldots = 0,$$

i.e. the algebraical sum of the resolved parts of the forces in any direction is zero, and the particle is therefore in equilibrium.

120.* The material systems, to which the following propositions refer, are either single rigid bodies, or systems of rigid bodies, connected in such a manner by means of inextensible strings, smooth joints, &c., that the motion of one of them determines that of all.

Prop. *A material system as above, under the action of a system of external and internal forces, is in equilibrium, provided that for every indefinitely small virtual displacement whatever, which does not violate the geometrical conditions, the algebraical sum of the virtual work done by the external forces is of an order higher than the displacement.*

For if the system be not in equilibrium, its motion is a definite one, which does not break the geometrical conditions: now we can conceive a number of smooth surfaces or inextensible strings to be so arranged, that they do not interfere with the actual motion of the system, but yet render it the only motion possible. If this be done, the whole system can be fixed by fixing one of the moving points in it, and this can be effected by applying to it a force (F) of sufficient magnitude, in the direction opposite to the point's motion, since that is the only direction in which the particle can move.

The system is now in equilibrium under the action of the original forces, the new forces of constraint and the force F. If then any virtual displacement be given to the system the algebraical sum of the virtual work done by these forces is zero: let the displacement be the one which actually takes place, when F is not applied. In this case the work done by the new forces of constraint is zero (Art. 117) and by hypothesis the virtual work done by the original forces alone is zero also. Hence the virtual work done by F is also zero. But as the point on which F acts is one of the moving points of the system, its displacement is of the first order, so that F must be zero, i. e. the system is in equilibrium without F, and without the new forces of constraint.

121.* *Prop. A material system as above, under the action of external and internal forces, will, if held in any position, and then let go, move at first so that the algebraical sum of the work done by the external forces is positive, provided the position is not one of equilibrium.*

If the system is not in equilibrium, we can, as before, arrange a system of smooth surfaces, so that the actual motion is the only one possible, and this again can be entirely prevented by applying at one of the moving particles, A, say, of the system, a force, of sufficient magnitude F, in a direction opposite to that of A's motion. The system being now in equilibrium we see as before, choosing the virtual displacement that which actually takes place, that the algebraical sum of the virtual work done by the original forces and F is zero.

But it is obvious, since A moves in the direction opposite to that of the force F, the work done by F is initially negative; and therefore the algebraical sum of the virtual work done by the original forces is initially positive. The algebraical sum of the work actually done by the original forces is therefore at first positive.

122. *Def.* The equilibrium of a body is said to be *stable*, when, on moving it slightly from its position of equilibrium, it returns to it; if it moves still further away from this position, its equilibrium is *unstable*. If the body remain in equilibrium, the equilibrium is *neutral*.

If a small ring slide on a smooth circular wire placed in a vertical plane, it will be found by experiment that there are two positions of equilibrium, one, the stable one, at the lowest point of the wire, and towards which it will readily return if moved away from it: the other, the unstable one, at the highest point, away from which it will move if disturbed ever so slightly, and in which it is found practically almost impossible to keep it. A uniform sphere resting on a smooth horizontal plane is an instance of a body in neutral equilibrium; for if it be rolled out of its position along the plane, it will neither return, nor, unless a velocity be given to it, move further away.

The positions of stable and unstable equilibrium of a body succeed one another alternately, i.e. there cannot be two positions of stable equilibrium without one of unstable equilibrium between them, and vice versâ. For a position of stable equilibrium is one to which the body tends to move when placed near it, so that if there are two such positions of a body, there must be a position between them such that, if the body be placed on one side of it, it will tend to move towards one of the above positions, and if placed on the other side, towards the other: i.e. there is a position of unstable equilibrium between them. Similarly we can shew, that there is a position of stable equilibrium between every two positions of unstable equilibrium.

123.* Prop. *When the only external forces acting on the material system of the last two propositions are the weights of the different particles which compose it, and the forces due to the geometrical constraints, such as the reactions of smooth surfaces and the tensions of inextensible strings, the system is in a position of stable or unstable equilibrium, according as its centre of mass is at a maximum or minimum depth consistent with the geometrical conditions of constraint.*

For the work done by gravity during any displacement is the algebraical sum of the products of the weight of each particle into its *vertical* displacement (the positive sign being given to the displacement when it is downwards, the negative when upwards): and this again is equal to the product of the weight of the whole system into the *vertical* displacement of its centre of mass. Also, during any displacement of the system, consistent with the geometrical conditions, gravity is the only force which does work. We have seen (Art. 121), that when not in equilibrium the system moves so that the work done by the forces is initially positive, i.e. in this case, so that the centre of mass moves downwards. Hence the system always tends to move initially, so that its centre of mass moves towards the adjacent position at a maximum depth and away from the adjacent position at a minimum depth; these positions succeed one another alternately, and it is clear that the former are positions of stable, and the latter of unstable equilibrium.

124.* The cases considered above divide themselves into two classes: one, in which the centre of mass of the system is constrained to move along a certain *curve*, so that in any position, it is only free to move in *two* directions, opposite to one another; the other, in which the centre of mass is constrained to move on a certain *surface*, so that in any position, it is free to move in *any* direction in a certain plane, the tangent plane to the surface at the point. A rod with its ends compelled to move along fixed wires is an illustration of the first class, one placed inside a bowl is an illustration of the second class.

If there is a position of the system, such that for all possible small displacements from it, the depth of the centre of mass is *diminished*, that position will be a position of *absolutely stable* equilibrium; on the other hand, a position, such that for all possible small displacements from it, the depth of the centre of mass is *increased*, is one of *absolutely unstable* equilibrium. A point then on the locus of the centre of mass, where the tangent line or plane is horizontal, corresponds to a position of equilibrium; if the tangent line or plane is below the adjacent points of the curve or surface, the corresponding position of the system is one of absolutely stable equilibrium, if it is above the adjacent points, one of absolutely unstable equilibrium.

If the locus of the centre of mass is a curve, it may be that there is a point on it, such that the tangent at it is horizontal, and cuts the curve there, i.e. the adjacent part of the curve on one side is above the tangent and that on the other side below it: in other words there may be a point of inflexion at which the tangent is horizontal. Such a position of the centre of mass corresponds to a position of equilibrium of the system, a position from which a displacement in one direction brings about a tendency to return to it, in the other direction, a tendency to recede still further from it.

Again, when the locus of the centre of mass is a surface, the shape of the latter may be that of a saddle, or that of the ground at the top of a pass between two mountains; in this case a tangent plane to the ground at the top of the pass is horizontal, and has part of the surface above it and part below it. This position of the centre of mass corresponds to a position of equilibrium of the system, which is unstable for displacements of the centre of mass in the plane containing the tangent to the path over the pass, and stable for displacements in the plane at right angles to it.

125.* A body BAC rests on a rough fixed body DAE, the surfaces near the point of contact A being spherical: it is required to determine whether, for displacements made by *rolling only*, BAC is in stable or unstable equilibrium.

Let o, O be the centres of the spherical surfaces: we suppose that the common normal oAO is vertical. G the centre of mass of BAC will be situate in Ao.

Let BAC be displaced by rolling through a small angle so that it comes into the position $B'A'C'$, G' and o' being the new positions of G and o, P the point of contact of the two surfaces.

Let $oA = r$, $OA = R$, $AG = h$, $\angle AOP = a$, and $\angle A'o'P = \beta$.

The angle $o'A'$ makes with $Ao = \angle A'o'P + \angle PoA = a + \beta$,

∵ the arc $AP =$ the arc $A'P$, $Ra = r\beta$.

Since the weight of $B'A'C'$ acts through G', the equilibrium is stable or unstable, according as the vertical line through G' lies to the left or right of P, i.e. according as the horizontal distance from A of

G' is $<$ or $>$ that of P.

But the horizontal distance of G' from A

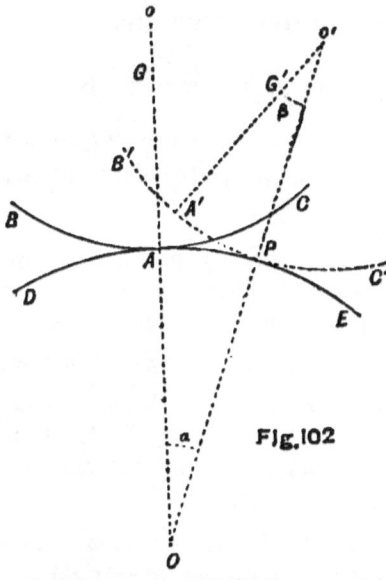

Fig. 102

= the horizontal distance of G' from A' since (Art. 117) AA' is of the second order, i.e.

$= A'G' \sin(\alpha + \beta)$.

The horizontal distance of P from $A = R \sin \alpha$,

∴ the equilibrium is stable or unstable according as $A'G' \sin(\alpha + \beta)$ is < or > $R \sin \alpha$,

according as h is < or > $\dfrac{R\alpha}{\alpha+\beta}$ as α and β are small,

according as h is < or > $\dfrac{Rr}{R+r}$,

according as $\dfrac{1}{h}$ is > or < $\dfrac{1}{R}+\dfrac{1}{r}$.

If the concavity of either surface be turned the other way we shall obtain the same result as before, except that the sign of the corresponding radius will be changed. If either surface be plane, its radius is of course infinity.

Cor. The above results hold for any curved surfaces, if R and r represent the radii of curvature of the sections made by the plane of displacement.

If $\dfrac{1}{h} = \dfrac{1}{r} + \dfrac{1}{R}$, the equilibrium is said to be *critical*, and we must proceed to a higher degree of approximation in order to determine whether the equilibrium is really *stable* or *unstable*.

Ex. 1. A body made up of a cone and a hemisphere having a common base, rests with the axis vertical on a rough horizontal table: determine the greatest height of the cone in order that the equilibrium may be stable. *Ans*: Height of cone $= \sqrt{3}$. radius of base.

Ex. 2. A prolate spheroid rests with its axis horizontal on a rough horizontal plane; shew that for rolling displacements in its equatorial plane the equilibrium is *neutral*, and for displacements in the vertical plane containing the axis, it is *stable*.

Ex. 3. A right circular cylinder of radius r rests with its axis horizontal on a fixed rough sphere (radius $R > r$): shew that the equilibrium is stable or unstable, according as the plane in which the displacement takes place makes with the vertical one containing the axis of the cylinder an angle $<$ or $> \cos^{-1}\sqrt{(r/R)}$.

Ex. 4. A prolate hemispheroid rests with its vertex on a rough horizontal plane, prove that the equilibrium is stable or unstable according as the eccentricity of the generating ellipse is less or greater than $\sqrt{(3/8)}$.

126.* The material systems for which we have proved the preceding propositions have been either single rigid bodies, or rigid bodies connected in such a way that the position of one determines the positions of the others. We can however easily extend them to include the case of a system of rigid bodies so connected, that it is necessary to know the positions of a number of the bodies in order to know those of all.

Prop. *If the algebraical sum of the virtual work done by the external forces be zero for all possible small virtual displacements consistent with the geometrical conditions, the above material system is in equilibrium.*

For if it is not, it will have a definite motion consistent with the geometrical conditions, and without interfering with the actual motions of the bodies we can so arrange a number of smooth surfaces or inextensible strings, that these actual motions are the only ones possible: they need not however all take place, i.e. several of the bodies may move without all the others doing so, and fixing one of them will not of necessity fix all. In this case we can reduce the whole system to rest by fixing one moving point in each of the bodies, and this can be done by applying forces P, Q, R, &c. of sufficient magnitude in the directions opposite to the actual motions of these points respectively. Now the whole system is in equilibrium under the action of the original forces, the new forces introduced by the smooth fixed surfaces &c., and the forces P, Q, R, &c.: therefore, if the virtual displacements chosen be the actual ones, the algebraical sum of the virtual work done by the original forces and P, Q, R, &c. will be zero, because the virtual work done by the reactions of the smooth surfaces is zero. But it is obvious that the work done by each of the forces P, Q, R, &c. is negative, since the particle on which it acts moves in the direction opposite to that of the force: the algebraical sum of the work done by the original forces is therefore positive, which is inconsistent with its being zero, as it is by supposition. Hence each of the forces P, Q, R, &c. is zero, and the system is in equilibrium; and as the smooth surfaces or inextensible strings do not interfere with the actual motion in any way, their removal will not upset the equilibrium of the system.

Cor. We see from the foregoing, that when such a material system as the above is not in equilibrium in a particular position, under the action of given external forces, it will, if placed in that position and then released, move so that at first the algebraical sum of the work done by the external forces is positive. By reasoning as in Art. 123 the proposition there proved can be extended to the case of the material systems we have just been considering. Among the systems we can include a mass of liquid, or a heavy inextensible flexible string.

It may be observed that the propositions of Arts. 120, 121 apply to all systems of bodies the internal forces among which can do no work, so long as the geometrical conditions are not violated: they will not however apply to those systems in which the internal forces are capable of doing work; for instance, systems in which the pressures of compressible fluids, the tensions of elastic strings, and the actions of rough surfaces are included among the internal forces. On the other hand there is no restriction on the nature of the external forces: they may consist of frictions, or the tensions of elastic strings, without affecting the validity of these propositions.

127.* In a precisely similar way to that used in Art. 121, we can prove the much more general proposition still, that *if any material system whatsoever, under the action of any system of forces, be placed in any position and then released, it will, if not in equilibrium, move at first so that the work done by all the forces, internal as well as external, is positive.*

128.* *Def.* When the forces, internal as well as external, acting on a material system are such, that the algebraical sum of the work done by them, as the configuration of the system changes, depends only on the initial and final configurations and not on the paths the different bodies take, they are said to form a *Conservative* system of forces.

Def. If any material system is acted on by a conservative system of forces, the algebraical sum of the work done by these forces, as the configuration of the system changes from any other to some standard configuration, is termed the *Potential Energy* of the system corresponding to the former configuration. It is generally convenient to take the standard configuration such that the potential energy for every other configuration which is practically considered is positive.

129.* *Prop.* When any material system is acted on by a conservative system of forces, it is in a position of *stable* equilibrium when its potential energy has a *minimum* value, and in a position of *unstable* equilibrium when its potential energy has a *maximum* value.

We have seen (Art. 127), that when the system is placed in any position, except one of equilibrium, and then released, it will move so that the algebraical sum of the work done by the forces is initially *positive*, i.e. it will move so as to *diminish* its potential energy. Hence it will move towards a position of minimum, and away from one of maximum potential energy. The positions of *maximum* potential energy then are positions of *unstable* equilibrium and those of *minimum* potential energy of *stable* equilibrium.

The proposition proved in Art. 123 is a particular case of this theorem.

130. *Recapitulation.* We began by shewing that if a particle be in equilibrium, the total virtual work done by the forces acting on it, during any virtual displacement whatever, is zero. The same theorem is therefore true for any system of particles, when the *internal* as well as the *external* forces are taken into consideration; but if the virtual displacement is a small quantity of the first order, and the system of particles form a rigid body, and also in certain other cases, it was shewn that the total virtual work done by the *external forces alone* is a small quantity of the second order.

The converse theorem was then shewn to hold for single rigid bodies, and also for a system of rigid bodies, connected in certain ways. Also such a system will, if placed in any position, and then released, move so that the total virtual work done by the external forces during the initial small displacement is positive, unless the position is one of equilibrium. Hence followed the principle, that such a material system, when gravity is the only *active* force, is in stable or unstable equilibrium, according as its centre of mass is at a maximum or minimum depth consistent with the geometrical conditions. Similarly followed the more general theorem, that for any material system under the action of any *conservative* system of forces, stable positions are positions of minimum, and unstable positions of maximum, potential energy.

ILLUSTRATIVE EXAMPLES.

Ex. 1. Find the amount of work done in stretching an elastic string.

Let a be the natural length of the string, λ its modulus of elasticity; let x be the extension of the string.

Let the extension x be divided into n, an indefinitely large number, equal parts. When the length of the string is $a + \dfrac{rx}{n}$, the tension is $\lambda \cdot \dfrac{r}{n} \cdot \dfrac{x}{a}$, and therefore the work done in stretching it to $a + \dfrac{r+1}{n} x$ lies between $\quad \lambda \cdot \dfrac{r}{n} \cdot \dfrac{x}{a} \cdot \dfrac{x}{n}$, and $\lambda \cdot \dfrac{r+1}{n} \cdot \dfrac{x}{a} \cdot \dfrac{x}{n}$,

i.e. the total work done in stretching the string to the length x

$$= \frac{\lambda x^2}{a} \cdot L_{n=\infty}^{t} \frac{1+2+3+\ldots(n-1)}{n^2} = \frac{\lambda x^2}{2a}.$$

Hence the work done in increasing the extension from y to x is $\lambda (x^2 - y^2)/2a$.

Ex. 2. Shew that the power necessary to move a cylinder of radius r and weight W up a plane inclined at angle a to the horizon by a crowbar of length l inclined at an angle β to the horizon is

$$\frac{Wr}{l} \cdot \frac{\sin a}{1 + \cos(a + \beta)}.$$

Let O be the point where the axis of the cylinder intersects the vertical plane containing the crowbar AB; C the point where the same plane meets the generating line in contact with the inclined plane. Let P be the force, which applied at B at right angles to AB will maintain equilibrium.

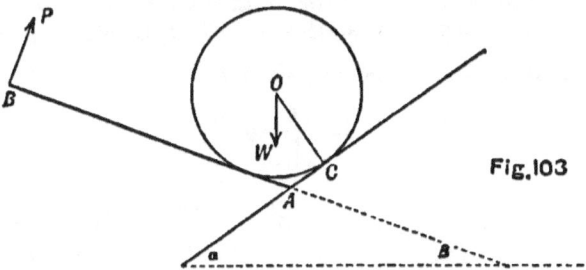

Fig. 103

Let the virtual displacement be for AB to turn through a small angle θ, so that its inclination to the horizon becomes $\beta + \theta$.

By Art. 117, the actions between the cylinder and crowbar and between each and the plane do not enter into the equation of virtual work.

The vertical height of O above A is $AC \sin \alpha + OC \cos \alpha$, i.e.

$$= r \{\sin \alpha \tan \tfrac{1}{2}(\alpha + \beta) + \cos \alpha\} = r \cos \tfrac{1}{2}(\alpha - \beta)/\cos \tfrac{1}{2}(\alpha + \beta).$$

$\therefore$ neglecting the weight of the crowbar, the equation of virtual work is

$$Pl\theta - Wr \left(\frac{\cos \frac{\alpha - \beta - \theta}{2}}{\cos \frac{\alpha + \beta + \theta}{2}} - \frac{\cos \frac{\alpha - \beta}{2}}{\cos \frac{\alpha + \beta}{2}} \right) = 0.$$

$$\therefore P = \frac{Wr}{l} \cdot \frac{\sin \alpha}{\cos \frac{\alpha + \beta}{2} \cos \frac{\alpha + \beta + \theta}{2}} \cdot \frac{\sin \frac{\theta}{2}}{\theta}$$

$$= \frac{Wr}{l} \cdot \frac{\sin \alpha}{2 \cos^2 \frac{\alpha + \beta}{2}} \cdot \frac{\cos \frac{\alpha + \beta}{2}}{\cos \frac{\alpha + \beta + \theta}{2}} \cdot \frac{\sin \frac{\theta}{2}}{\frac{\theta}{2}}$$

$$= \frac{Wr}{l} \cdot \frac{\sin \alpha}{1 + \cos (\alpha + \beta)},$$

since θ^2 and higher powers are neglected.

Ex. 3. A straight uniform rod has smooth small rings attached to its extremities, one of which slides on a fixed vertical straight wire, and the other on a fixed wire in the shape of a parabola whose latus rectum equals twice the length of the rod, and whose axis coincides with the straight wire: prove that in the position of equilibrium (stable when the vertex is upwards) the rod will be inclined at an angle of 60° to the vertical. Which is the position of stable equilibrium when the vertex is downwards?

Let PQ be the rod, P being the point on the parabola. Let θ be its inclination to the vertical, $2a$ its length, and G its middle point.

Draw PN perpendicular to the axis AN.

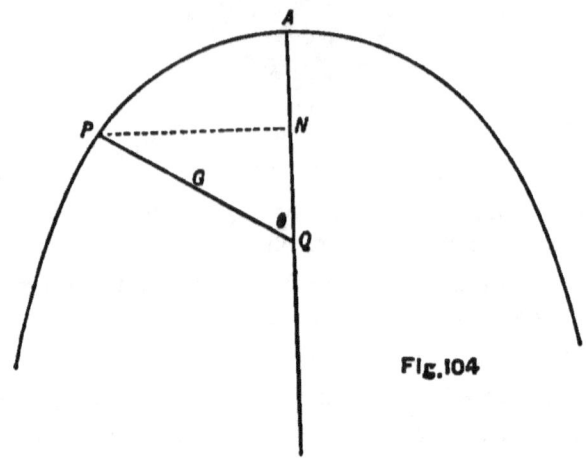

Fig. 104

(1) When the axis is upwards, the depth of G below A
$$= AN + PG \cos\theta$$
$$= \frac{PN^2}{4a} + a\cos\theta = \frac{4a^2 \sin^2\theta}{4a} + a\cos\theta$$
$$= a(1 + \cos\theta - \cos^2\theta).$$

The positions of equilibrium are given by the maximum and minimum values of this expression,
$$1 + \cos\theta - \cos^2\theta = \tfrac{5}{4} - (\tfrac{1}{2} - \cos\theta)^2,$$
i.e. is a maximum when $\cos\theta = \tfrac{1}{2}$ or when $\theta = 60°$.

It is clearly a minimum when $\theta = 0$.

Hence $\theta = 60°$ corresponds to a position of stable equilibrium, and $\theta = 0$ to one of unstable equilibrium.

(2) When the vertex is downwards $\theta = 0$ corresponds to the position of stable, and $\theta = 60°$ to the position of unstable equilibrium.

Ex. 4. Two smooth rods which intersect at an angle 2α are placed so that they are equally inclined to the vertical, and the line bisecting the angle between them is inclined at an angle β to the vertical. Prove that a spherical ball of radius a will be in a position of unstable equilibrium,

if the distance of its points of contact with the rods from the intersection of the rods be
$$\frac{a \cot a \cos \beta}{\sqrt{(1 - \cos^2 a \sin^2 \beta)}}.$$

Let O be the centre of the sphere, O' of the circle in which it is intersected by the plane of the rods. Let B, C be the points where the rods AB, AC touch the sphere: θ the angle CO' or BO' subtends at O.

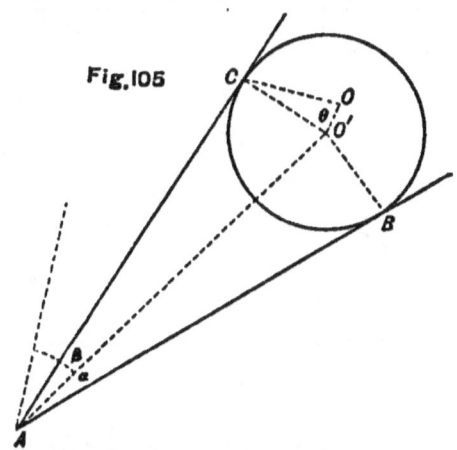

Fig. 105

Then $OO' = a \cos \theta$, $CO' = a \sin \theta$, $AO' = a \sin \theta \operatorname{cosec} a$.

The vertical height (h) of O above A
$$= AO' \cos \beta + OO' \sin \beta,$$
$$= a (\sin \theta \operatorname{cosec} a \cos \beta + \cos \theta \sin \beta).$$

Let $\cos \beta \operatorname{cosec} a = r \cos \phi$, and $\sin \beta = r \sin \phi$:
then $\qquad h = ar (\cos \phi \sin \theta + \cos \theta \sin \phi) = ar \sin (\theta + \phi).$

Now the sphere is in a position of unstable equilibrium when h is a maximum, i.e. when $\theta + \phi = \frac{1}{2}\pi$, i.e. when $\theta = \cot^{-1}(\sin a \tan \beta)$.

But $\qquad AB = O'B \cot a = a \cot a \sin \theta,$
$$\therefore AB = \frac{a \cot a}{\sqrt{(1 + \sin^2 a \tan^2 \beta)}} = \frac{a \cot a \cos \beta}{\sqrt{(1 - \cos^2 a \sin^2 \beta)}}.$$

Ex. 5. A trestle composed of four jointed bars which form a crossed parallelogram $ABCD$, the alternate bars being equal, is placed in a vertical position, and is stiffened by a cord connecting the lower corners B and D, which rest on a frictionless horizontal plane. A platform is supported on AC, and is loaded in any manner. Shew that the stress in the connecting cord is independent of the distribution of this load W, and is equal to $W / (\tan \alpha + \tan \beta)$, where α and β are the inclinations of AB and AD to BD.

Draw AH perpendicular to DB: let $AH = h$, $AB = a$, and $AD = b$: let T be the tension of cord DB.

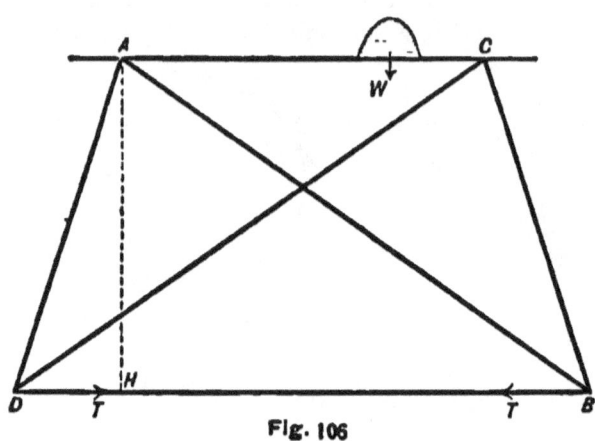

Fig. 106

$$AH = AB \sin ABD = AD \sin ADH,$$
$$\therefore \; h = a \sin \alpha = b \sin \beta,$$
$$DB = HB + HD = a \cos \alpha + b \cos \beta.$$

Let such a virtual displacement be given to the system that the joints at A, B, C, D are unbroken, while the height AH is diminished by a small quantity x, and the length DB is increased by y. Let in consequence α become $\alpha - \alpha'$, and β, $\beta - \beta'$.

Then
$$h - x = a \sin (\alpha - \alpha') = b \sin (\beta - \beta'),$$
and
$$DB + y = a \cos (\alpha - \alpha') + b \cos (\beta - \beta'),$$
$\therefore$
$$x = a \{\sin \alpha - \sin (\alpha - \alpha')\} = 2a \cos (\alpha - \tfrac{1}{2}\alpha') \sin \tfrac{1}{2}\alpha'$$
$$= a\alpha' \cos \alpha \text{ ultimately.}$$

Similarly
$$x = b\beta' \cos \beta.$$
$$y = a \{\cos (\alpha - \alpha') - \cos \alpha\} + b \{\cos (\beta - \beta') - \cos \beta\}$$
$$= 2a \sin \tfrac{1}{2}\alpha' \sin (\alpha - \tfrac{1}{2}\alpha') + 2b \sin \tfrac{1}{2}\beta' \sin (\beta - \tfrac{1}{2}\beta'),$$
$$= a\alpha' \sin \alpha + b\beta' \sin \beta \text{ ultimately}$$
$$= x (\tan \alpha + \tan \beta).$$

VIRTUAL WORK. 221

Since none of the connections are broken except the cord DB, the only forces which do work are the weight W which, however distributed, descends through a distance x, and the tension T, which does negative work. (It is assumed that the bars themselves are without weight.)

Hence the equation of virtual work is

$$Wx - Ty = 0,$$

$$\therefore T = Wx/y = W/(\tan \alpha + \tan \beta).$$

Ex. 6. Four uniform thin heavy rods are freely jointed together at their extremities so as to form a parallelogram, and two opposite angular points of the frame so formed are connected by a light inextensible string; the system is suspended by another string attached to one of the same angular points: compare the tensions of the strings.

Let A be the point from which the frame is suspended, B the diagonally opposite point to which the string is attached.

The centre of mass, G, is the middle point of AB, which is therefore vertical. Let the virtual displacement be such that B is moved vertically downwards through a small distance x without any separation of the rods at the joints.

By Art. 117, the only forces which occur in the equation of virtual work are T, the tension of the string AB, and W the weight of the four rods.

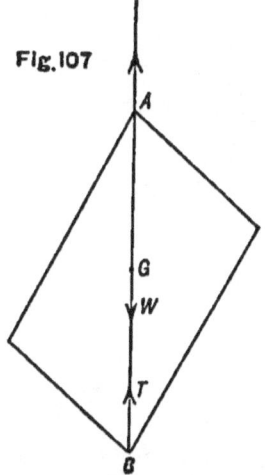

Fig. 107

The equation is $Tx - W \cdot \tfrac{1}{2}x = 0$;

$$\therefore T = \tfrac{1}{2}W,$$

i.e. the tension of the string AB is half that of the one which supports the whole framework.

The same reasoning will enable us to prove that the same relation holds when the framework of rods forms the edges of a parallelepiped, or any figure, such that the centre of mass is always the middle point of the diagonal along which the string lies.

Ex. 7. Three equal particles, each of weight W, are fastened to an endless elastic string without weight, so as to be at equal distances from each other. The whole is then laid on a smooth sphere so that the string lies unstretched along a horizontal small circle of the sphere whose radius is $\tfrac{4}{5}$ that of the sphere. Prove that the particles will be in equilibrium when the lines joining them subtend angles of $60°$ at the centre, the modulus of elasticity of the string being $\tfrac{3}{4}\sqrt{3}\,W$.

Let O be the centre of the sphere; A, B, C the positions of the particles when in equilibrium. Let H be the point where the vertical through O meets the plane ABC, which is from symmetry horizontal. Let $r =$ radius of the sphere. Let ϕ be the angle which AB, BC, or CA subtends at O, and θ the angle which OA, OB, or OC makes with OH.

$$\angle AHB = \angle AHC = \angle BHC = \tfrac{2}{3}\pi,$$

$$HA = r \sin \theta.$$

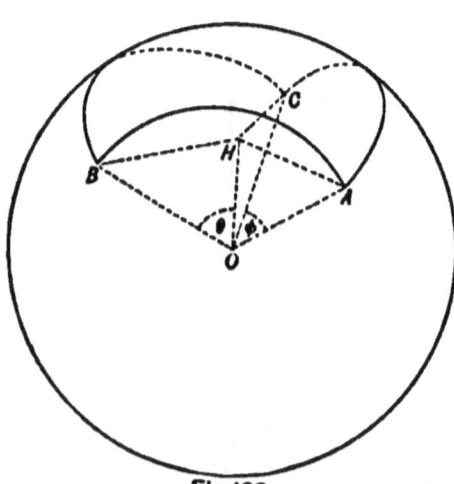

Fig. 108.

VIRTUAL WORK. 223

$$\therefore AB = 2r \sin \theta \sin \tfrac{1}{3}\pi = \sqrt{3} \cdot r \sin \theta;$$

$$\therefore 2 \sin \tfrac{1}{2}\phi = \sqrt{3} \sin \theta \quad \ldots\ldots\ldots\ldots\ldots\ldots\ldots\ldots(1).$$

The original length of the string was $\tfrac{3}{4}\pi r$; when stretched it is $3r\phi$. T, the tension, therefore

$$= \frac{3r\phi - \tfrac{3}{4}\pi r}{\tfrac{3}{4}\pi r} \cdot \frac{3W}{2\sqrt{2}}, \quad = \frac{3W}{2\sqrt{2}} \cdot \frac{4\phi - \pi}{\pi}.$$

Let the virtual displacement be such that all three particles descend through equal small distances, the consequent small increments in θ and ϕ being x and y. The equation of virtual work is then

$$3Wr\{\cos\theta - \cos(\theta + x)\} - 3Try = 0,$$

$$\therefore 2W \sin(\theta + \tfrac{1}{2}x) \sin \tfrac{1}{2}x - \frac{3W}{2\sqrt{2}} \cdot \frac{4\phi - \pi}{\pi} y = 0,$$

$$\therefore \pi x \sin \theta = \tfrac{3}{4}\sqrt{2}(4\phi - \pi) y \quad \ldots\ldots\ldots\ldots\ldots\ldots(2).$$

From (1) $\quad 2 \sin \tfrac{1}{2}(\phi + y) = \sqrt{3} \sin(\theta + x) \quad \ldots\ldots\ldots\ldots\ldots(3);$

$\therefore$ subtracting (1) from (3) we have

$$2 \cos(\tfrac{1}{2}\phi + \tfrac{1}{4}y) \sin \tfrac{1}{4}y = \sqrt{3} \cos(\theta + \tfrac{1}{2}x) \sin \tfrac{1}{2}x,$$

$$\therefore y \cos \tfrac{1}{2}\phi = x \sqrt{3} \cos \theta \quad \ldots\ldots\ldots\ldots\ldots\ldots(4);$$

eliminating the ratio $x : y$ between (2) and (4),

$$\tfrac{3}{4}\sqrt{6}(4\phi - \pi) \cos \theta = \pi \sin \theta \cos \tfrac{1}{2}\phi,$$

substituting from (1)

$$\tfrac{3}{4}\sqrt{6}(4\phi - \pi)(3 - 4\sin^2 \tfrac{1}{2}\phi)^{\tfrac{1}{2}} = 2\pi \cos \tfrac{1}{2}\phi \sin \tfrac{1}{2}\phi = \pi \sin \phi.$$

This equation is satisfied by putting $\phi = \tfrac{1}{4}\pi$, and substituting in (1) we get a consistent value for θ; this value of ϕ therefore corresponds to a position of equilibrium.

Ex. 8. A small ring sliding on a smooth elliptic wire, whose axis is vertical is connected by elastic strings with each focus, the modulus of elasticity is half the weight of the ring, and either string is just unstretched when the ring is as near as possible to the corresponding focus. Shew that in the unsymmetrical position of equilibrium the distance of the ring from the upper focus is equal to the distance of the centre from either directrix. Determine the nature of the equilibrium in the different positions.

Let P be any position of the ring, S', A' the upper focus and vertex, S, A the lower; let $SP = r$. Let W be the weight of the ring.

224 STATICS.

Let the system have its standard configuration (Art. 128) when the ring is at A. In that case the potential energy of the system when the ring is at P is the total work done by the forces as the ring moves from P to A.

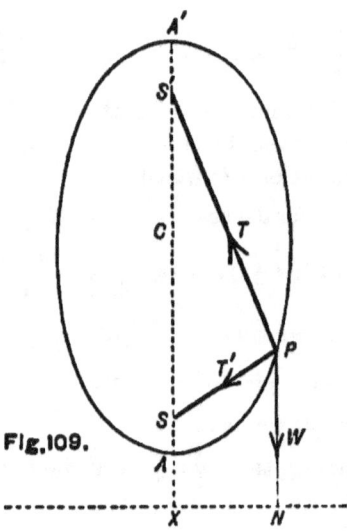

Fig. 109.

Draw PN perpendicular to the directrix NX. The potential energy of the system, when the ring is at P, is (Ex. 1, page 216),

$$W(PN - AX) + \frac{W}{2} \cdot \frac{SP^2 - SA^2}{2SA} - \frac{W}{2} \cdot \frac{S'A^2 - S'P^2}{2SA}$$

$$= W \left\{ \frac{SP - a(1-e)}{e} + \frac{SP^2 + S'P^2 - a^2(1-e)^2 - a^2(1+e)^2}{4a(1-e)} \right\}$$

$$= W \left\{ \frac{r - a(1-e)}{e} + \frac{r^2 - 2ar + a^2(1-e^2)}{2a(1-e)} \right\}$$

$$= W \frac{er^2 + 2a(1-2e)r - a^2(1-e)(2-3e-e^2)}{2ae(1-e)}$$

$$= W \frac{\{er - (2e-1)a\}^2 - a^2(1 - 2e - e^2 + 2e^3 + e^4)}{2ae^2(1-e)}.$$

The minimum value of this expression corresponds to $er = (2e-1)a$, i.e. $S'P = a/e = CX$; the maximum values correspond to the greatest and least values of SP, i.e. occur when P coincides with A and A'.

Hence the stable position of the ring is where its distance from $S' = CX$, and the unstable positions are A and A'. This supposes that there is a point P on the ellipse such that $S'P = a/e$, which will not be the case unless e be less than a certain quantity. If there is no such point, it is easily seen that A' is the position of unstable and A of stable equilibrium.

EXAMPLES.

1. A heavy beam AB is movable freely about the end A which is fixed; an elastic string is attached to A, passes through a fixed ring C vertically above A and is fastened to B: AC is equal to AB. Find the position of equilibrium. If the natural length of the string be AC, discuss the problem in the case when its modulus of elasticity is $>$, $=$, or $<$ half the weight of the beam.

2. A rhombus is composed of four equal rods jointed at their extremities. Two opposite corners are connected by an elastic string whose natural length is $a\sqrt{2}$, a being the length of each rod, and the system stands in a vertical plane with one of the corners on a horizontal table. Find the angle between the rods.

3. A solid homogeneous hemisphere of radius a and weight W rests in apparently neutral equilibrium on the top of a fixed sphere of radius b. Prove that $5a = 3b$. A weight P is now fastened to a point in the rim of the hemisphere. Prove that if $55P = 18W$, the hemisphere can still rest in apparently neutral equilibrium in contact with the highest point of the sphere.

4. Two heavy rings slide on a fixed smooth parabolic wire whose axis is horizontal, and the rings are connected by a string which passes over a smooth peg at the focus. Prove that in the position of equilibrium the depths of the rings below the axis of the parabola are proportional to their weights. Is the equilibrium stable or unstable?

5. A prolate spheroid rests upon another equal and similar fixed spheroid, the point of contact being on the equatorial plane of each, their major axes being horizontal and at right angles to each other. Prove that the equilibrium will be stable for a displacement in a plane through either axis, if the upper spheroid be loaded at its lowest point with a weight bearing to its own weight a ratio greater than the duplicate ratio of its least and greatest diameters.

6. A uniform rod AB of length $2a$ is freely movable about A: a smooth ring of weight P slides on the rod and has attached to it a fine string which passes over a pulley at a height b vertically above A and supports a weight Q hanging freely; find the position of equilibrium of the system.

7. A cylinder rests in equilibrium with the centre of its base on the highest point of a fixed and perfectly rough sphere. The altitude and diameter of the base of the cylinder are each equal in length to a quadrant of a great circle of the sphere. Find the greatest angle through which the cylinder may be made to rock without falling off.

8. A wire in the form of an ellipse, whose semi-axes are a and b, is placed with its minor axis vertical. A light string of length a on which slides a ring of weight W has one end fastened to the centre, and the other to a ring of weight W', which slides on the wire. Shew that, if there is no friction, there will be equilibrium if W' is anywhere on the upper half of the ellipse, and $b/a = W'/(W+2W')$.

9. Two particles are connected by a fine inextensible string and can move freely in a smooth cycloidal tube whose vertex is upwards, the string passing over the vertex. Prove that in equilibrium the arcual distances of the particles from the vertex must be inversely as their masses.

10. Five equal rods each of length a are hanged together and placed on a smooth horizontal table, one of the angular points being joined to the two opposite angles by two equal strings of length $2c$: a horizontal force P acts at the middle point of each rod in an outward direction perpendicular to its length. Find the tensions of the strings: and shew that the action at each of the joints where there is no string is parallel to the nearest string and is equal to

$$\tfrac{1}{2} Pa \, (a^2 - c^2)^{-\tfrac{1}{2}}.$$

11. A parallelogram composed of jointed rods, each of length a and weight P, is hung up by one angle, and inside it is placed a circular disc of radius b and weight W. Prove that there will be equilibrium, when θ the inclination of the rods to the vertical is given by the equation

$$2a \, (W+2P) \sin^3 \theta = Wb \cos \theta.$$

12. A lamina in the form of a rhombus made up of two equilateral triangles rests with its plane vertical between two smooth pegs in the same horizontal plane at a distance apart equal to a quarter of the longer diagonal: prove that either a side or a diagonal of the rhombus must be

vertical, and that the stable position is that in which a diagonal is vertical.

13. A parallelogram $ABCD$ formed of four uniform rods freely jointed at the corners has the side AB fixed horizontally, and the frame hangs in a vertical plane with the joint A attached by a light string of length l to the opposite joint C: AC is the shorter diagonal and a the acute angle of the parallelogram: shew that the tension of the string is $\frac{Wl}{2a}\cot a$, where a is the length of the fixed side and W the weight of the four rods.

14. A surface rests in contact with a perfectly rough fixed surface, the common normal at the point of contact making an angle a with the vertical: prove that the equilibrium is stable or unstable, according as the distance of the centre of mass from the point of contact is less or greater than
$$\frac{\cos a}{1/\rho + 1/\rho'},$$
where ρ, ρ' are the radii of the surfaces, supposed spherical at the point of contact.

15. A heavy body in the shape of a paraboloid of revolution placed on a rough horizontal plane, has its C. G. at the critical height: determine this height, and find the real nature of the equilibrium.

16. A thin straight rod is suspended by a fine inextensible string fastened to it at the two ends and passing over a fixed smooth peg. If the centre of gravity of the rod is not at its middle point, determine whether the equilibrium is stable or not.

17. A uniform rod of length c rests with one end on a smooth elliptic arc whose major axis is horizontal and with the other on a smooth vertical plane at a distance h from the centre of the ellipse: prove that, if θ be the angle which the rod makes with the horizon and $2a$, $2b$, the axes of the ellipse, $2b \tan\theta = a \tan\phi$, where $a\cos\phi + h = c\cos\theta$.

18. Two elastic strings are fastened at a fixed point P and pass through fixed smooth rings A and B such that PA, PB are the natural lengths of the respective strings: the other ends of the strings are fastened to C and D, two points of a rigid lamina which is movable in its plane about a fixed point O. If A and B are in the same plane as the lamina and if the angles COA, DOB are supplementary and the system is in equilibrium, prove that the equilibrium will be neutral.

19. Twelve equal uniform rods form a cube having universal joints at each of its angles; shew that, if it be suspended by one of its angles, and be prevented from collapsing by three rods without weight forming the diagonals not passing through the point of suspension, the thrusts along the three rods will be half the weight of the framework.

20. Two equal rods rigidly fastened at right angles to each other are placed over an ellipse whose plane is vertical and major axis horizontal; find the least length of the rods that the equilibrium may be stable.

21. A smooth fixed sphere supports a zone of very small equal smooth spherical particles and the whole is prevented from slipping off the sphere by an elastic ring occupying a horizontal circle of angular radius a, shew that in the position of equilibrium the tension of the band is T, where $2\pi T = W \tan a$, and W is the whole weight of the ring and particles together.

22. A uniform elliptic hoop is weighted at an extremity of its major axis by a weight equal to that of itself: shew that if it be placed on a smooth horizontal plane with its plane vertical, it will have two or four positions of equilibrium according as its eccentricity is less or greater than $\frac{1}{2}\sqrt{2}$. What is the nature of the equilibrium in the several positions?

23. Two similar uniform straight rods of lengths $2a$, $2b$ rigidly united at their ends at an angle a rest over two smooth pegs in the same horizontal plane: prove that the angle which the rod $2a$ makes with the vertical is given by the equation

$$c(a+b)\sin(2\theta - a) = a^2 \sin a \sin \theta - b^2 \sin a \sin(a-\theta),$$

c being the distance between the pegs.

24. Three equal and in every way similar uniform rods AB, BC, CD freely jointed at B and C, have small smooth weightless rings attached to them at A and D: the rings slide on a smooth parabolic wire whose axis is vertical and vertex upwards, and whose latus rectum is half the length of the three rods: prove that in the position of equilibrium, the inclination (θ) of AB or CD to the vertical is given by the equation

$$\cos \theta - \sin \theta + \sin 2\theta = 0.$$

Is the equilibrium stable or unstable?

25. A number of uniform thin rods, all equal and similar, are freely jointed together at their middle points, so that they form the generators of a right circular cone, symmetrically placed about the axis. Within

the cone thus formed is placed a smooth sphere, and round the rods a smooth thin ring of the same weight and radius as the sphere. The whole is placed on a smooth horizontal plane, so that the ring is below and the sphere above the vertex of the cone; prove that the semi-vertical angle (θ) of the cone in one position of equilibrium is given by

$$(P+W) a (2 \sin \theta + \sin 2\theta) = Pr,$$

where W is the weight of the rods, P that of the sphere and ring together, $2a$ the length of each rod, and r the radius of the ring or sphere. Determine the stress on any rod at the joint.

26. A, B, C, D are four fixed points in the same horizontal plane, at the corners of a square whose semi-diagonal is b. A', B', C', D' are the corresponding corners of a square plate of weight W and semi-diagonal a. Four equal cords join AA', BB', CC', DD'. When the plate is hanging by the cords the distance between $A'B'C'D'$ and $ABCD$ is k. Shew that if a couple L about a vertical axis be applied to the plate so that it is turned through an angle θ, then

$$L = Wab \sin \theta / \sqrt{(k^2 - 4ab \sin^2 \tfrac{1}{2}\theta)}.$$

27. Two equal equilateral triangular laminæ each of weight W and freely jointed together at their vertices, are placed with their bases on a smooth horizontal table, and have their base angular points connected by two inextensible strings, one of which is equal in length ($2a$) to a side of either triangle. Shew that the tension of the other string ($2b$) is equal to

$$\tfrac{1}{3} Wa^2 (3a-b)^{-\tfrac{3}{2}} (2a-b)^{\tfrac{1}{2}}.$$

28. Two small rings, each of weight P, slide on a smooth circular wire (radius r) in a vertical plane, and are connected by a string of length $2a$ ($< 2r$) on which slide a ring of weight Q. Shew that when the string is vertical, the corresponding position is one of unstable equilibrium, and that the stable position of equilibrium is when Q is at a distance from the centre of the wire

$$= \sqrt{\{(r^2 - a^2) P / (P+Q)\}}.$$

29. A pyramidal plug is made to fit symmetrically into an equilateral triangular hole whose side is a and plane horizontal. Prove that, to retain it in the hole with its axis vertical so that its section by the plane of the hole is an equilateral triangle whose side is c, a couple must be applied of moment $Wh \left(\dfrac{4c^2}{a^2} - 1 \right)^{\tfrac{1}{2}}$, where W is the weight of the plug and h is the depth of the vertex in this position.

CHAPTER VII.

MACHINES.

131. It is frequently desirable that we should be able to counteract one force by another, differing from it in magnitude, point of application, or direction, or in all three. To enable us to do this we employ machines more or less complicated.

In Statics we suppose the machine to be in equilibrium under the action of the forces due to the geometrical conditions of constraint, the force at our disposal generally called the *Power*, and the force which we wish to counteract, generally called the *Resistance* or the *Weight*.

It is found practically that, when the power is just on the point of overcoming the weight, other resistances are called into play, owing chiefly to the friction between the different parts of the machine, and the imperfect flexibility of ropes: all these resistances oppose the power, so that the latter has to be greater than would be necessary, were the machine a perfect one. If the weight were on the point of overcoming the power, these resistances would assist the latter. It is usual to call the resistance or weight, which it is the object of the machine to enable us to overcome, the *useful* resistance, while the other resistances are called *wasteful* resistances. When we take these latter into consideration, we shall suppose that motion is just about to take place, and that the power is overcoming the useful resistance.

If motion just occurs, the work done by the power will equal that done against both the useful resistance and the wasteful ones; the former part of the work is termed *useful* and the latter *lost* work.

132. *Def.* When motion just takes place in a machine, the ratio of the *useful work* done to the *whole work* done in the same indefinitely short time is called the *Efficiency* of the machine. It is of course desirable to have the efficiency as near unity as possible.

Let P denote the power, W the useful resistance, and W' the wasteful resistance.

If P move its point of application through a small distance s, and in consequence the work done against W be w, and that done against W' be w', we have from the principle of virtual work,

$$Ps = w + w'$$

the efficiency then is $w/(w+w')$.

Let P_0 be the force which would just move W were there no wasteful resistance, then $P_0 s = w$ by the principle of virtual work. Hence the efficiency $= P_0 s/Ps = P_0/P$, or the efficiency is the ratio of the power, which would just move the weight were there no wasteful resistance, to the actual power required.

Unless otherwise stated, we shall suppose the machines perfect ones, i.e. with efficiency equal to unity.

133. The simple machines are,—the *Lever*, the *Wheel and Axle*, the *Pulley*, the *Inclined Plane*, the *Screw* and the *Wedge*. The principle of the wheel and axle is the same as that of the lever, and the screw and wedge are identical in principle with the inclined plane.

134. *The Lever.* This is a rigid rod, straight or curved, and free to turn about a fixed axis, which is called the *fulcrum*. The two parts into which the rod is divided by the fulcrum are called *arms*.

Levers are usually classified as follows. In the lever of the *first* class, the fulcrum is between the power and the weight: a poker where the bar of the grate is used as the fulcrum, and a pair of scissors are instances of it. In the *second* class, the weight is between the fulcrum and the power, as in a wheelbarrow, where the point of the wheel in contact with the ground is the fulcrum, or in an oar, where the blade in contact with the water is the fulcrum, and the resistance is applied at the rowlock. In levers of the *third* class, of which a pair of shears and the human arm are examples, the power is between the fulcrum and the weight.

135. *The condition of equilibrium of a Lever.* As in Art. 74 we can shew that the necessary and sufficient condition of equilibrium of any body whatsoever, which is free to turn about a fixed axis, and under the action of any number of forces, is, that the algebraical sum of the moments of the forces about the fixed axis be zero. In the case of the simplest form of the lever the forces are generally only two, the power and the weight, acting in one plane, so that the condition of equilibrium becomes that the moment of P about the fulcrum should be numerically equal but of opposite sign to that of W.

This condition may also be easily found by the Principle of Virtual Work.

136. *To determine the pressure on the fulcrum when the Lever is in equilibrium.*

Since the action of the fulcrum together with the power and the weight keeps the lever in equilibrium, the reaction on the fulcrum is obviously the resultant of the power and the weight. If, however, the lines of action of P and W are not in one plane, they do not reduce to a single resultant, and the pressure on the fulcrum is not a single force.

We shall assume that the lines of action of P and W are in one plane.

MACHINES.

When the power and the weight are parallel, the reaction (R) of the fulcrum (F) is parallel to each of them; and in a lever of the first class, $R = P + W$,

of the second class, $R = W - P$,

of the third class, $R = P - W$.

When the lines of action of the power and weight are not parallel but meet in C, let A, B be their respective

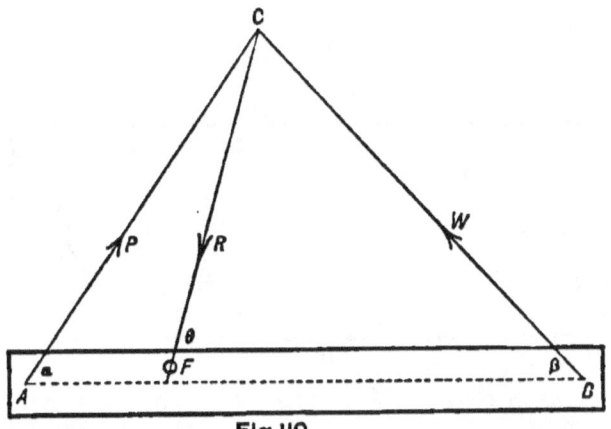

Fig. 110.

points of application, α, β the angles, which their lines of action make with AB.

It is required to find the magnitude of R, and the angle (θ) its direction makes with AB.

Since the lever is in equilibrium under the action of the three forces, R's line of action passes through C.

Also (Art. 18) $\sin ACF : \sin BCF = W : P$;

$\therefore \sin(\theta - \alpha) : \sin(\pi - \theta - \beta) = W : P$;

$\therefore \dfrac{\sin\theta \cos\alpha - \cos\theta \sin\alpha}{\sin\theta \cos\beta + \cos\theta \sin\beta} = \dfrac{W}{P}$;

$\therefore \tan\theta = \dfrac{W \sin\beta + P \sin\alpha}{P \cos\alpha - W \cos\beta}$.

Also $R^2 = P^2 + Q^2 + 2PQ \cos ACB$;

$\therefore R = \sqrt{\{P^2 + Q^2 - 2PQ \cos(\alpha + \beta)\}}$.

137. *To find the relation between the Power and the Weight in a rough Lever, when the Power is on the point of moving the Weight.*

Let A, B be the points of application of the power (P) and the weight (W) respectively: let their lines of action

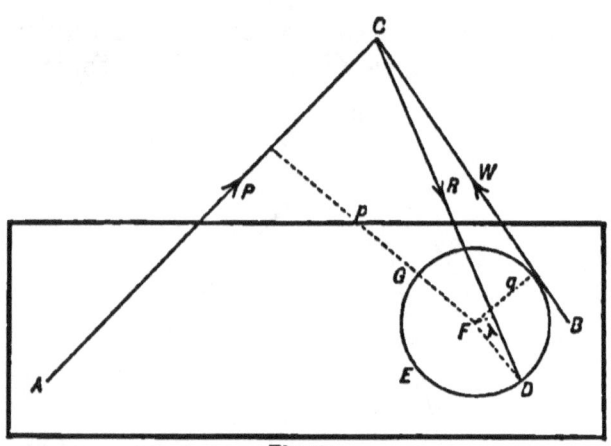

Fig. 111

meet in C at an angle θ. The fulcrum is a rough solid cylinder, which passes through a cylindrical hole in the lever, just so much bigger in diameter that there is contact along one generating line only.

Let the plane ABC, which we assume to be perpendicular to the axis of either cylinder, cut the hole in the circle DEG of radius r and centre F, D being the point where the line of contact meets the circle. Join DC, then the reaction of the fulcrum (R) acts along CD. Also, since the lever is on the point of turning round F in the direction in which P tends to turn it, the reaction R will make with the normal FD an angle equal to λ, the angle of friction, and on the side which enables it to assist W.

Let p, q be the perpendiculars from F on the lines of action of P, W respectively.

Since R counteracts P and W,

$$R^2 = P^2 + W^2 + 2PW \cos \theta.$$

Also, by taking moments about F, we have

$$Pp = Wq + Rr \sin \lambda$$
$$= Wq + r \sin \lambda \cdot \sqrt{(P^2 + W^2 + 2PW \cos \theta)} \ldots (1).$$

If P and W are in the same direction this becomes

$$P(p - r \sin \lambda) = W(q + r \sin \lambda).$$

If P could only just balance W, or, in other words, were W on the point of moving P, the relation would be

$$Pp = Wq - r \sin \lambda \cdot \sqrt{(P^2 + W^2 + 2PW \cos \theta)}.$$

138. *To find the efficiency (E) of the rough Lever.*

Let P_0 be the power just required to move W, when the fulcrum is perfectly smooth.

Then $\qquad P_0 p = Wq.$

But the efficiency, $\quad E = \dfrac{P_0}{P} = \dfrac{Wq}{Pp}.$

Therefore from (1) we have

$$1 = E + \frac{r \sin \lambda}{p} \cdot \sqrt{\left(1 + \frac{p^2 E^2}{q^2} + 2\frac{pE}{q} \cos \theta\right)};$$

$$\therefore pq(1 - E) = r \sin \lambda \cdot \sqrt{(q^2 + p^2 E^2 + 2pq E \cos \theta)},$$

which gives us E.

If P and W are in the same direction E becomes

$$q(p - r \sin \lambda)/\{p(q + r \sin \lambda)\}.$$

139. *The Wheel and Axle.* This machine consists of a cylinder a (the wheel) with a groove cut round the circumference, and a cylinder b of smaller radius (the axle). The two form one rigid body and have a common horizontal axis cc', at the ends of which are two pivots c and c', resting in fixed sockets so that the whole can turn about this axis.

The power P is applied tangentially at the circumference of the wheel, generally by means of a rope, while

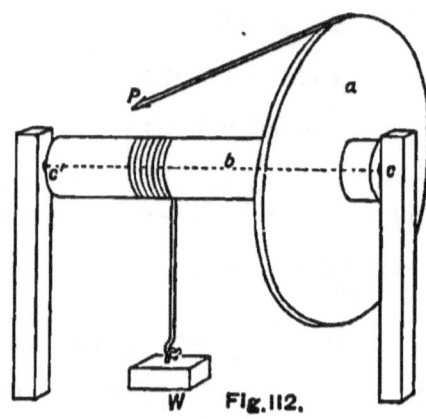

Fig. 112.

the weight is suspended by a rope which is wound round the axle so that it tends to turn the machine in the opposite direction to the power.

The apparatus for drawing a bucket of water out of a well is frequently a machine of this kind, the power being applied by means of a handle attached to the wheel instead of by a rope. A windlass for hauling up an anchor on board ship is a modification of the wheel and axle, in which the common axis is vertical, and the power is applied at the end of poles which project from the wheel so as to form radii produced.

Condition of Equilibrium. The Wheel and Axle, as before stated, is a kind of lever, and we can shew as in Art. 74 that the condition of equilibrium is that the moment of P about the axis should be equal and opposite to the moment of W, i.e. that

$P \times$ the radius of the wheel $= W \times$ radius of the axle.

If the ropes be of considerable thickness, the tension of each may be supposed to act along its axis or central line, so that the condition of equilibrium becomes

$P \times$ (rad. of wheel $+ \frac{1}{2}$ thickness of P's rope)
$= W$ (rad. of axle $+ \frac{1}{2}$ thickness of W's rope).

Rigidity of Ropes. We have hitherto supposed, that the ropes are *perfectly flexible*, i.e. that they offer no resistance to being bent. As a matter of fact when a rope is wound *on* to a drum, pulley or axle, it does offer a resistance, though none is offered when it is wound *off*. From experiments made on new dry ropes and tarred ones Coulomb has deduced the following empirical results.

If a rope whose tension is T, is on the point of being wound on to a drum, the effect of the *rigidity* of the rope is the same as would be produced by increasing T by a certain amount T'. This amount T' is most simply expressed by the formula

$$T' = \frac{a + bT}{R'},$$

where a and b are constants depending on the rope, and R' is the *effective* radius of the drum, i.e. its actual radius + half the thickness of the rope.

What is the precise meaning of the above statement?

The force T exerted by the rope cannot be greater than itself; how can the rigidity of the rope increase the effect of T? As it cannot alter the magnitude and direction of T, it clearly can only increase T's effect by causing it to act at a greater distance from the drum than would be the case, if the flexibility were perfect. This is precisely what we should be led to expect from à priori considerations. Where the rope is being *wound on*, those fibres furthest from the drum will be stretched more than those nearer, and will therefore exert the greatest tensions—hence the resultant tension will act *further from* the drum than the central axis of the rope. At the point where the rope is being *wound off*, there seems no reason why one fibre should exert a greater tension than another so that the resultant tension will act along the central axis.

Where must T act in order to have as great a moment about the axis of the axle or drum, as $T + \dfrac{a + bT}{R'}$ acting along the central axis of the rope? If the distance of the line of application of T from the central axis be x, we must have

$$T(R' + x) = \left(T + \frac{a + bT}{R'}\right) R',$$

$$\therefore \quad x = \frac{a}{T} + b,$$

x clearly cannot be greater than ½ the thickness of the rope.

238   STATICS.

By using the method of Art. 137, the following results may be obtained.

If the Power P applied to a wheel and axle of weight w, by means of a rope of thickness t_1, be about to raise a weight W which is suspended by a rope of thickness t_2, the relation between P and W is

$$P(R+\tfrac{1}{2}t_1-\rho\sin\lambda)=W(r+\tfrac{1}{2}t_2+\rho\sin\lambda)+w\rho\sin\lambda,$$

where R is the radius of the wheel, r that of the axle, ρ that of the pivots about which the whole turns and λ the angle of friction between the pivots and their bearings.

If the rigidity of the rope to which W is attached be taken into account, the relation becomes

$$P(R+\tfrac{1}{2}t_1-\rho\sin\lambda)=\left(W+\frac{a+bW}{r+\tfrac{1}{2}t_2}\right)(r+\tfrac{1}{2}t_2+\rho\sin\lambda)+w\rho\sin\lambda,$$

where a and b are the constants mentioned above.

In both these cases, P is applied vertically downwards.

Ex. 1. Four sailors, each exerting a force of 112 lbs., can just raise an anchor by means of a capstan whose radius is 1 ft. 2 in. and whose spokes are 8 ft. long (measured from the axis). Find the weight of the anchor. *Ans.* $1\tfrac{3}{5}$ tons.

Ex. 2. If the length of each of a pair of sculls be 8 ft. 6 in., and the distance from the hand to the rowlock be 2 ft. 3 in., find the resultant force on the boat when the sculler pulls each scull with a force of 25 lbs., assuming that the blade does not move through the water. *Ans.* 18 lbs.

Ex. 3. A fly-wheel 10 ft. in radius weighs 15 tons, its axle is 6 in. in radius and revolves in bearings between which and it the coefficient of friction is ·2: find the smallest weight which, hung from a band round the circumference of the wheel, will just turn it. *Ans.* 333 lbs. nearly.

Ex. 4. Find the efficiency of a 'wheel and axle,' which weighs 50 lbs. and turns on pivots of $\tfrac{1}{2}$ in. radius, and coefficient of friction ·1, when the Power acts vertically downwards, the radii of the wheel and axle are 2 ft. 8 in. and 5 in. respectively, and the weight to be raised is 500 lbs.

Ans. ·987.

Ex. 5. Find the efficiency of the machine described in Ex. 4, everything remaining the same, except that the thickness of the ropes and their rigidity are to be taken into account; the ropes are ·8 in. thick, and the constants a, b (Art. 139), are 8·6 and ·18 respectively, provided that, in using the empirical formula of Art. 139, R' is expressed in inches, and T in lbs. *Ans.* ·952.

MACHINES. 239

140. *The Pulley.* A pulley-block consists of two plates or sheaves connected by an axle about which a circular disc, with a groove cut in its circumference, can turn. Rigidly connected with the axle is a hook to which a string can be attached so as to support the pulley, or by means of which the pulley can support a weight. Sometimes there are several discs, either turning about the same axle or placed one below another; they then form *double*, *treble*, &c. blocks. A rope passes along the groove in the circumference of the disc, and, as the latter is supposed smooth, the tension of the rope will be the same on both sides the pulley.

When the block is fixed, the pulley is said to be *fixed;* otherwise it is called a *movable* pulley.

If a *fixed* pulley be used to enable us to overcome resistance, the only object gained by the use of the pulley is that the force applied is enabled to counteract a force in a *different* direction, though not of greater magnitude.

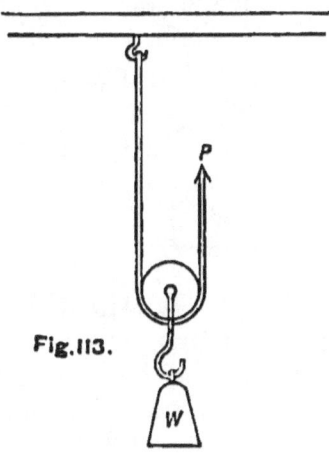

Fig. 113.

When a *single movable* pulley is used, the weight W is attached to the block, and the power P is applied at one end of a rope which passes under the disc of the pulley, the other end of the rope being fastened to a fixed point.

It is obvious, when the strings are parallel, that $W = 2P$, and when they are not parallel, but each makes an angle θ with the vertical, that $W = 2P \cos \theta$.

141. There are three systems of pulleys usually described in text-books.

In the *First System*, the weight is attached to the lowest pulley, which is supported by a rope, one end of which is

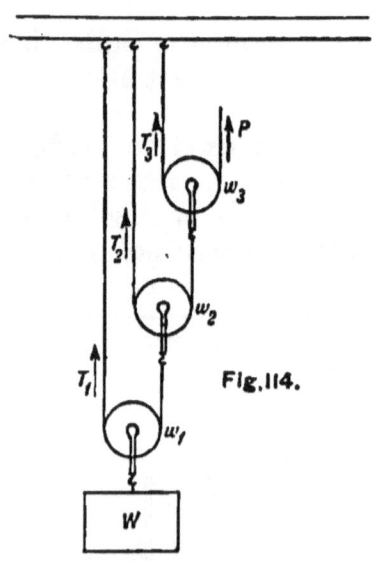

Fig. 114.

attached to a fixed beam, and the other end to the next pulley above, which is in turn supported in a similar way: the power is applied at the end of the rope supporting the highest pulley. The portions of the ropes not in contact with the pulleys are vertical.

Condition of Equilibrium. Let there be n pulleys, whose weights in order from the lowest are $w_1, w_2, \ldots w_n$, and let the tensions of the ropes supporting them be $T_1, T_2, \ldots T_n$.

Then from the equilibrium of the different pulleys, we have

$$2T_1 = W + w_1 \quad\quad\quad\quad\quad (1),$$
$$2T_2 = T_1 + w_2 \quad\quad\quad\quad\quad (2),$$
$$2T_3 = T_2 + w_3 \quad\quad\quad\quad\quad (3),$$
$$\ldots = \ldots\ldots\ldots\ldots\ldots\ldots\ldots$$
$$2T_n = T_{n-1} + w_n \quad\quad\quad\quad\quad (n),$$

also
$$P = T_n \quad\quad\quad\quad\quad (n+1).$$

Multiplying equations $(2), (3)\ldots(n+1)$ by $2, 2^2, 3^3,\ldots 2^n$ respectively, and adding, we have

$$2^n \cdot P = W + w_1 + 2w_2 + 2^2 w_3 + \ldots 2^{n-1} w_n.$$

If the pulleys be without weight, this equation reduces to $2^n \cdot P = W$.

We can deduce the same equation by the principle of virtual work.

Let the virtual displacement be the one which would actually be produced by moving the end of the rope to which P is applied through a small distance x in P's direction. By this, the uppermost pulley would be raised through a height $\frac{x}{2}$, the next lower pulley through a height $\frac{x}{2^2}$, and so on, the lowest pulley and weight being raised through a height $\frac{x}{2^n}$.

During this displacement, the actions of the fixed points to which the ends of the different ropes are attached do no work, nor is any done by the internal forces of the system (Art. 117). The equation of virtual work then is

$$P \cdot x - w_n \cdot \frac{x}{2} - w_{n-1} \cdot \frac{x}{2^2} - \ldots - (W + w_1)\frac{x}{2^n} = 0,$$

i.e. $2^n P = W + w_1 + 2w_2 + 2^2 w_3 + \ldots 2^{n-1} w_n.$

G.

142. In the *Second System* there are two pulley-blocks, the upper of which is fixed, and the lower movable: a rope passes over one of the discs of the upper block and under one of the lower block alternately, the radii of the different discs being such that the portions of the rope not in contact with a pulley are vertical, or nearly so. One end of the rope is attached to one of the

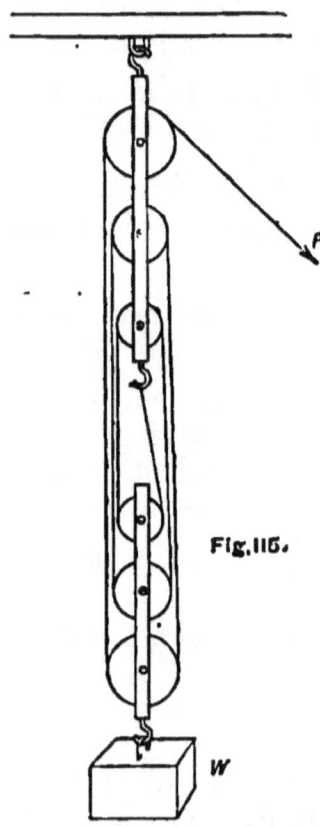

Fig. 116.

two blocks, and at the other end the power is applied. The weight is attached to the lower block.

Condition of Equilibrium. Let W be the weight to be raised, including that of the lower block: let P be the power which just raises it: then the tension of the rope is

MACHINES.

P throughout, and if there be n strings coming from the lower block, the total force exerted by them is nP, and we must have $W = nP$.

143. In the *Third System*, the uppermost pulley is fixed: each pulley has a rope passing over it, with one end attached to the weight and the other to the pulley next below. The power is applied at the end of the string passing over the lowest pulley.

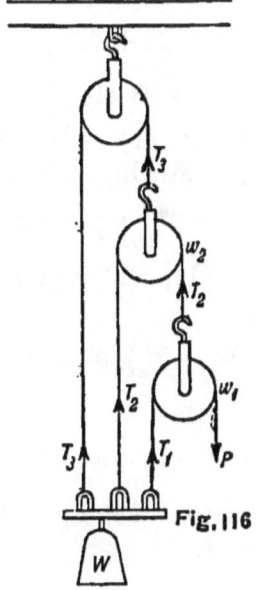

Fig. 116

It should be observed that the third system is obtained by inverting the first, the beam becoming the weight, and vice versâ.

Condition of Equilibrium. In investigating the relation between the weight W and the power P which will support it, we shall suppose the portions of the ropes not in contact with the pulleys to be vertical, and the ropes to be without weight.

Let there be n pulleys, including the fixed one, and let $w_1, w_2, w_3 \ldots w_{n-1}$ be the weights of the movable ones beginning with the lowest; let $T_1, T_2 \ldots T_n$ be the tensions of the ropes passing over them.

Since each pulley is in equilibrium, we have

$$T_2 = 2T_1 + w_1 \ldots \ldots \ldots \ldots (1),$$
$$T_3 = 2T_2 + w_2 \ldots \ldots \ldots \ldots (2),$$
$$\ldots = \ldots \ldots$$
$$T_n = 2T_{n-1} + w_{n-1} \ldots \ldots (n-1),$$

from the equilibrium of the weight

$$T_1 + T_2 + T_3 + \ldots T_n = W \ldots \ldots \ldots (n),$$

also $\quad T_1 = P$.

Multiplying equations (1), (2)…$(n-1)$ by 2^{n-1}, 2^{n-2}, $2^{n-3}\ldots 2$ respectively, and adding, we have

$$2T_n = 2^{n-1} w_1 + 2^{n-2} w_2 + \ldots 2w_{n-1} + 2^n P.$$

Adding equations (1), (2)…$(n-1)$, and employing equation (n), we have

$$W - P = 2(W - T_n) + w_1 + w_2 + \ldots w_{n-1}.$$

Eliminating T_n, we have

$$W = P(2^n - 1) + w_1(2^{n-1} - 1) + w_2(2^{n-2} - 1) + \ldots w_{n-1}(2 - 1).$$

To deduce the relation between W and P from the principle of virtual work.

Let the weight W be supposed moved vertically downwards through a small distance x. Then the highest movable pulley, w_{n-1}, will be raised through a height x: the next pulley below will be raised twice the height through which the highest is raised, together with the distance through which the weight descends, i.e. through a height $3x$. Similarly we can see that any pulley will rise through a height x, together with twice the distance through which the next pulley above rises. The distances therefore through which the weights $w_{n-1}, w_{n-2} \ldots w_1$, are respectively raised are $x, (2^2 - 1)x, (2^3 - 1)x, \ldots (2^{n-1} - 1)x$.

MACHINES. 245

Also the point of application of P will be moved vertically upwards through a distance $(2^n - 1)x$.

Hence the equation of virtual work is

$$Wx - w_{n-1}x - (2^2 - 1)w_{n-2}x - (2^3 - 1)w_{n-3}x$$
$$- (2^{n-1} - 1)w_1 x - (2^n - 1)Px = 0;$$
$$\therefore W = (2^n - 1)P + (2^{n-1} - 1)w_1 + (2^{n-2} - 1)w_2$$
$$+ \ldots (2^2 - 1)w_{n-2} + w_{n-1}.$$

We can take into account the friction between the axles of the pulleys and their bearings by the method of Art. 137, and also the rigidity of the ropes by that of Art. 139.

Ex. 1. If there are three movable pulleys arranged as in the first system, their weights beginning from the lowest being 9, 3, and 1 lbs. respectively, find what power will support a weight of 69 lbs. *Ans.* 11 lbs.

Ex. 2. If in the second system there are altogether nine pulleys and each pulley weigh one pound, what force will be required to support a weight of 86 lbs. ? *Ans.* 10 lbs.

Ex. 3. If the weight supported in the third system be 56 lbs., and each movable pulley, of which there are 3, weigh 1 lb., find the horizontal distance of the centre of mass of the weight from the centre of the fixed pulley, supposing the diameters of all the pulleys to be equal.
Ans. $\frac{9}{23}$ the radius of any pulley.

Ex. 4. If a weight P be on the point of lifting a weight Q by means of a rope to which P is attached, passing over a fixed pulley and under a movable one, to the latter of which Q is attached: find the relation between P and Q, assuming that the pulleys are exactly similar, and that the effects of friction, and rigidity, are small.

Ans. $P = \frac{1}{2}\{1 + \frac{3}{2}(2\rho \sin\phi + b)/r\}Q + \frac{3}{2}a/r$ where r is the radius of each pulley, ρ of the axle, ϕ the angle of friction, and a and b the constants determining the rigidity of the rope. (The ropes not in contact with the pulleys are supposed vertical.)

Ex. 5. Find the weight required to *lift* a weight of 70 lbs. by means of 2 movable pulleys arranged as in the first system, when each pulley weighs 1 lb., is of 3 in. radius, and has an axle $\frac{1}{4}$ in. radius, the coefficient of friction being ·075, and each rope ·6 in. thick, the constant of rigidity a being 3·6 and b ·125. *Ans.* 19·8 lbs. nearly.

Ex. 6. Find the force which must be applied vertically downwards in order to lift a weight by means of 4 pulleys arranged as in the second system, each of $2\frac{1}{4}$ in. radius, and turning about an axle of ·2 in. radius (coefficient of friction ·12), when the total weight of the lower block is 600 lbs., and the rope is ·25 in. thick. The constant of rigidity a is 1·8, and b, ·086. *Ans.* 172 lbs.

144. *The Inclined Plane.* A line in the plane perpendicular to its intersection with the horizontal plane is called a *line of greatest slope*, and the vertical plane containing this line the *principal* plane.

To find the condition of equilibrium on an inclined plane, where W is the weight and P the power.

(i) When the plane is *smooth*.

Let BAC be a section of the inclined plane made by a principal plane, BA being the line of greatest slope and

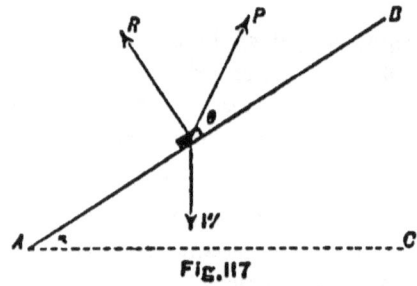

Fig. 117

AC being horizontal. Let α be the angle of inclination BAC.

The reaction R of the plane acts at right angles to the plane and therefore parallel to the plane BAC: the weight W acts parallel to this plane also, so that the power P must also act parallel to this plane. Let θ be the angle which P's line of action makes with AB measured up the plane, θ being positive when P's direction is above the plane.

By Art. 18,
$$P : W : R = \sin(W, R) : \sin(R, P) : \sin(P, W)$$
$$= \sin \alpha : \cos \theta : \cos(\alpha + \theta).$$

P clearly has its least value for a given value of W when $\theta = 0$.

(ii) When the plane is *rough*, and the direction of P is in the principal plane.

The total reaction R of the plane will act in the principal plane, since W and P do; its direction cannot

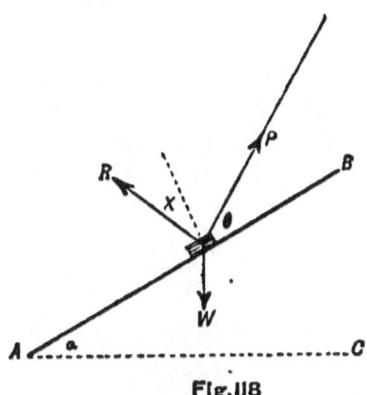

Fig. 118

make with the normal an angle greater than λ, the angle of friction, but may make any smaller angle.

Let λ' be the angle which R makes with the normal, λ' being measured towards the lower part of the plane.

We have then the equations
$$R^2 = P^2 + W^2 + 2PW \cos(\theta + \tfrac{1}{2}\pi + \alpha)$$
$$= P^2 + W^2 - 2PW \sin(\theta + \alpha),$$
and
$$\frac{P}{W} = \frac{\sin(\tfrac{1}{2}\pi - \lambda' + \tfrac{1}{2}\pi - \alpha)}{\sin(\lambda' + \tfrac{1}{2}\pi - \theta)} = \frac{\sin(\alpha + \lambda')}{\cos(\theta - \lambda')},$$

to determine R and λ'.

When P is just on the point of moving W so that the latter is just about to slip *up* the plane, the total reaction will make with the normal an angle λ on the side towards the lower part of the plane: in that case

$$P = W \cdot \frac{\sin(\alpha+\lambda)}{\cos(\theta-\lambda)}.$$

Also
$$R = W \cdot \frac{\cos(\alpha+\theta)}{\cos(\theta-\lambda)}.$$

The value of θ which will give the least value of P for a given value of W is λ; i.e. for P to be most effective it should make with the plane an angle equal to the angle of friction.

By changing the sign of λ in the preceding investigation we can obtain the value of P which will just prevent W from slipping *down* the plane, when the reaction R will make an angle λ on the other side of the normal. This gives us

$$P = W \cdot \frac{\sin(\alpha-\lambda)}{\cos(\theta+\lambda)},$$

and
$$R = W \cdot \frac{\cos(\alpha+\theta)}{\cos(\theta+\lambda)}.$$

(iii) When the plane is *rough* and P's direction does not lie in the principal plane.

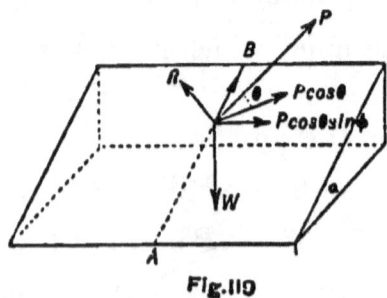

Fig. 110.

Let P make with the plane an angle θ, and let its resolved part along the plane make an angle ϕ with the line of greatest slope drawn up the plane. Let the reaction R of the plane be resolved into $R \cos \lambda'$ along the normal, and $R \sin \lambda'$ along the plane, the latter making an angle β with the line of greatest slope.

Resolving the forces at right angles to the plane, along the line of greatest slope, and in the plane at right angles to the line of greatest slope, we have

$P \sin \theta + R \cos \lambda' - W \cos \alpha = 0,$

$P \cos \theta \cos \phi + R \sin \lambda' \cos \beta - W \sin \alpha = 0,$

$P \cos \theta \sin \phi + R \sin \lambda' \sin \beta = 0.$

These equations are sufficient to determine R, λ' and β, when the other quantities are known.

If P be on the point of moving W, the reaction R makes with the normal an angle λ, so that writing λ for λ' in the above equations, they enable us to determine P, R and β, when the other quantities are known.

145. *The Screw.* A screw may be supposed constructed as follows :—

Let $aa'd'd$ be a solid right circular cylinder, and let $AA'D'D$ be a rectangle, whose breadth AA' is equal to

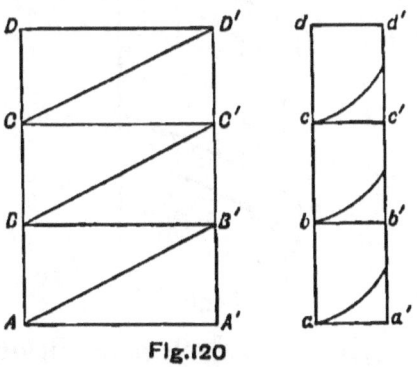

Fig. 120

the circumference of the cylinder. Draw BB', CC', DD', &c. parallel to AA' and at equal distances from one another: join AB', BC', CD'. Now let the rectangle AD'

be wrapped round the cylinder, so that the sides $ABCD$, $A'B'C'D'$ coincide with the generator ad, the points A, A' coinciding in a, B, B' in b, C, C' in c, and D, D' in d. The lines AB', BC', CD' will now form a continuous line going round the cylinder, called a *helix*. It is clear that in wrapping the rectangle round the cylinder, we have not altered the inclination to $A'B'$ of any of the lines AB', BC', CD', so that the helix everywhere makes the angle $B'AA'$ with the base of the cylinder. This angle is called the *pitch* of the helix; and it is equal to

$$\tan^{-1} \frac{AB}{AA'},$$

or $\tan^{-1} \dfrac{\text{distance between two consecutive coils } (ab)}{\text{circumference of the cylinder}}$.

Now imagine a solid figure to be generated by a small rectangle $abcd$ (fig. 121) which moves so that one side ad always coincides with a generating line, while a corner a describes the helix, and its plane always contains

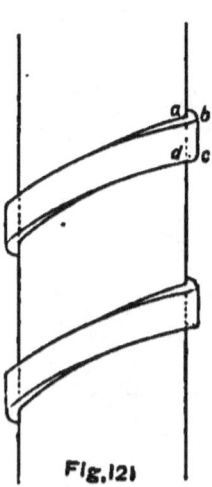

Fig. 121

the axis of the cylinder. Each point in ab will describe a helix, the pitch of the helix being smaller the further the point is from a: the distance between two consecutive coils will be the same for all, but the circumference of the

MACHINES. 251

cylinder round which any particular helix would wrap will be greater the further the generating point is from a.

This thread is a *square* one: an *angular* thread is generated by an isosceles triangle abc, whose plane always contains the axis of the cylinder, and whose base ab moves exactly as the side ad of the rectangle $abcd$ which generates the square thread.

The solid cylinder, together with the solid figure above described, form a solid screw, which works in a hollow cylinder of the same diameter as the solid one, and with a groove cut in it, which just fits the thread of the solid screw. The hollow screw is generally fixed in a support.

The screw is generally used as follows :—

The solid screw has at one end an arm at right angles to the axis: the power P is applied at the end of this arm

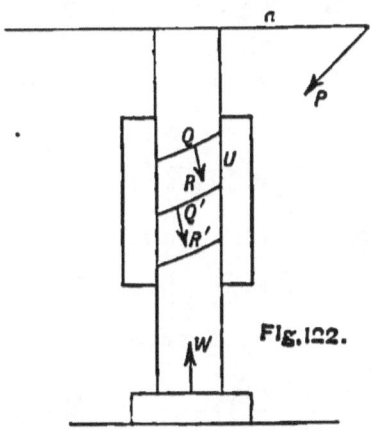

Fig. 122.

and perpendicular to it, so as to tend to turn the screw round and so move it in the direction of its axis, and thus produce pressure on any body situate at the end of the axis: the pressure which is thus overcome is called the Weight (W).

146. *To find the condition of equilibrium in a screw with a square thread.*

Let a be the length of the arm, at the end of which the power P acts: h the distance between two consecutive threads. The surface of the groove of the hollow screw will exert pressures perpendicular to the surface of the thread at a very large number of points. Let R be the resistance at one of these points Q, which is at a distance r from the axis of the screw: we have seen that the pitch (α) of the helix, which passes through this point, and has as axis the axis of the screw, is $\tan^{-1} h/2\pi r$. The direction of R is normal to the surface of the thread at Q, and therefore to any line in that surface, passing through Q. From the way in which the surface of the thread has been generated, R must be at right angles to the line from Q perpendicular to the axis of the screw; it must also be at right angles to the tangent to the helix through Q, i.e. it makes an angle α with the axis of the screw. We may resolve R into two components, one, $R \cos \alpha$, along the axis of the screw, and the other $R \sin \alpha$, perpendicular to the axis, and at a distance r from it. Similarly all the resistances, such as R, can be resolved in the same way.

Resolving along the axis, we have
$$W = \Sigma (R \cos \alpha).$$

Taking moments about the axis,
$$Pa = \Sigma (Rr \sin \alpha).$$

But $2\pi r \sin \alpha = h \cos \alpha$;

$$\therefore Pa = \Sigma \left(\frac{Rh \cos \alpha}{2\pi}\right) = \frac{h}{2\pi} \Sigma (R \cos \alpha) = \frac{Wh}{2\pi},$$

$$\therefore \frac{W}{P} = \frac{2\pi a}{h}$$

$$= \frac{\text{circumference of circle traced out by end of } P\text{'s arm}}{\text{distance between two consecutive threads}}.$$

We can easily deduce the same relation for a screw with any smooth thread, square or angular, from the

MACHINES. 253

principle of virtual work, by a method similar to that used in Art. 155.

Ex. A smooth screw makes three revolutions while it advances half an inch, find the power which must be applied at the extremity of an arm one foot long in order to produce a pressure of 144 lbs. *Ans.* ·32 lbs.

147. When the screw thread is not smooth, we can find the condition of equilibrium, if we assume that the breadth of the thread, i.e. the side ab (fig. 121) of the rectangle which is supposed to generate it, is very small. The pitch of the screw will be the same then at every point of the thread : let it be a.

Let us suppose that the power (P) is just on the point of moving the weight W, then the limiting friction is called into play at every point of contact of the thread with the groove, and acts in the direction in which it can most effectively oppose P, i.e. directly opposite to that in which the point of the thread is about to move: it makes then an angle $\frac{1}{2}\pi - a$, with the axis of the screw, and is perpendicular to the line drawn from its point of application at right angles to the axis. Let λ be the angle of friction, $R_1, R_2, R_3 \ldots$ the normal reactions at the different points of contact; r the distance of each of them from the axis of the screw.

Resolving along the axis of the screw, we have

$$W - \Sigma(R\cos a) + \Sigma(R\tan\lambda\sin a) = 0,$$

taking moments about the axis,

$$Pa - \Sigma(Rr\sin a) - \Sigma(Rr\tan\lambda\cos a) = 0.$$

But
$$r\sin a = \frac{h\cos a}{2\pi},$$

$$\therefore W = (\cos a - \sin a \tan\lambda)\Sigma(R) = \frac{\cos(a+\lambda)}{\cos\lambda}\Sigma(R),$$

and
$$Pa = \left(\frac{h\cos\alpha}{2\pi} + \frac{h\tan\lambda\cdot\cos^2\alpha}{2\pi\sin\alpha}\right)\Sigma(R)$$
$$= \frac{h\cot\alpha}{2\pi}\cdot\frac{\sin(\alpha+\lambda)}{\cos\lambda}\cdot\Sigma(R),$$
$$\therefore\ \frac{P}{W} = \frac{h}{2\pi a}\cdot\frac{\tan(\alpha+\lambda)}{\tan\alpha}.$$

When W is on the point of overcoming P, the relation becomes
$$\frac{P}{W} = \frac{h}{2\pi a}\cdot\frac{\tan(\alpha-\lambda)}{\tan\alpha}.$$

Since the power (P_0) which would just move W, when the screw is smooth, is $Wh/2\pi a$, the efficiency of the rough screw is $= \tan\alpha\cdot\cot(\alpha+\lambda)$.

Ex. If the screw in example on page 253 be rough, coefficient of friction ·08, and the radius of the axle be ½ in., find the power required to raise 144 lbs. *Ans.* ·8 lbs.

148. The *Wedge*, which is a solid prism, whose section is an isosceles triangle, and which is used to split wood, &c., by being driven in by blows of a hammer, is so essentially dynamical in principle, that we shall not discuss it here.

149. Besides being used as an instrument for multiplying Force, the Lever is also employed for weighing purposes: in one form it is known as the *Common Balance*.

This in its simplest form consists of a straight uniform beam AB, from the two ends of which scale-pans hang. The lever turns about a fulcrum C, which is situated above it in a short beam CD, which projects at right angles to AB, from its middle point D.

The substance to be weighed is placed in one scale-pan, and such weights in the other, that the beam is horizontal when in equilibrium.

An index needle is often attached at right angles to the beam. The needle traverses a graduated arc and shews when the beam is horizontal by pointing to the zero mark. In balances for accurate weighing the fulcrum

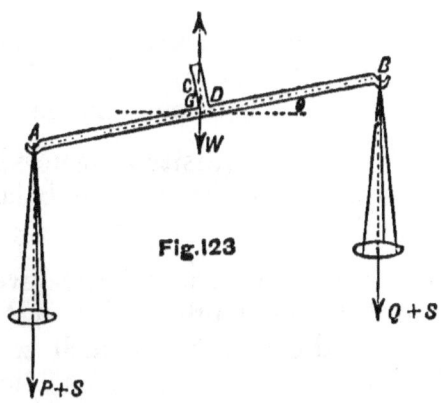

Fig. 123

is formed by the edge of a triangular prism of hardened steel (a knife edge), which rests on a plate of smooth agate. Hence (Art. 137) the effect of friction is rendered very small.

Fig. 123, like the other figures of the machines, is not intended as a *realistic* representation. It is assumed that the student is familiar with the actual forms of the simple machines.

150. The following are the requisites of a good Balance.

(1) It should be *true*, i.e. when loaded with equal weights, the beam should be horizontal. This requisite is obtained by making the scale-pans of equal weight, and the two arms exactly similar in every respect. We can easily *test* the *Truth* of a Balance by interchanging the weights, which keep the beam in equilibrium, when horizontal; if the beam settles again into a horizontal position, the weights are equal and the balance *true*, but not otherwise.

It is easy to make a balance approximately true, but almost impossible to make it absolutely so. When therefore very great accuracy is required, the method of *double weighing* is adopted. This enables us to determine the exact weight, however untrue the balance may be. It consists in first making the beam horizontal with the body whose weight is required in one scale-pan and sand or shot in the other: then the body is replaced by weights sufficient to keep the beam horizontal. It is clear that the weight of the body is that of the weights.

(2) A Balance should be *sensible*, i.e. when the weights differ by a small quantity the deviation of the beam from the horizontal should be easily perceptible.

To ascertain how this requisite is secured we must find the position of equilibrium when the balance is loaded with weights P and Q.

Let G be the centre of mass of the lever, not including the scale-pans, W its weight. Let $AB = 2a$, $CD = h$, $CG = k$. Let S be the weight of each scale-pan acting through A, B respectively. Let θ be the angle which AB makes with the horizon when P is placed in the scale-pan hanging from A and Q in the other.

By taking moments about C for the equilibrium of the beam, we have

$$(P + S)(a \cos \theta - h \sin \theta) - (Q + S)(a \cos \theta + h \sin \theta) - Wk \sin \theta = 0,$$

$$\therefore \tan \theta = \frac{(P - Q)a}{(P + Q + 2S)h + Wk}.$$

For a given value of $P - Q$, the sensibility will be the greater, the greater $\tan \theta$ is, and for a given value of θ, the sensibility is the greater, the smaller $P - Q$ is, so that we may take $\dfrac{\tan \theta}{P - Q}$ as a measure of the sensibility. Hence the second requisite is best obtained by making $\dfrac{a}{(P + Q + 2S)h + Wk}$ very large, i.e. by making a large in comparison with h and k.

In balances for very accurate weighing, C is in the line AB, so that h is zero, and the sensibility is proportional to a/Wk, which is independent of the load placed in the scale-pans. Also the sensibility is increased by making W smaller. In order to have a beam as long and as light as possible, consistent with sufficient strength, it is ordinarily made of not a single rod, but of a framework of rods.

(3) A good balance should be *stable*, i.e. it should readily return to its position of equilibrium, when moved from it, i.e. its time of oscillation about its position of equilibrium should be small. It is shewn in works on Rigid Dynamics that the time of oscillation is small when the arm a is small compared with h and k, so that the conditions of sensibility and stability are at variance one with another.

In making a balance, however, consideration is paid to the sort of weighing it is required for. In scientific measurements, where the greatest accuracy is desired, the third requisite is sacrificed to obtain the second; but for ordinary commercial purposes, where it is more necessary to save time than to be very accurate, the reverse is the case.

The stability is often measured by the sum of the moments of the forces which tend to bring back the beam into its position of equilibrium, but it is obvious that the time required to do this, and therefore the stability, will depend on the mass to be moved and on its shape, as well.

151. *The Common Steelyard.* This is a lever used as a balance, in which the necessity of keeping a number

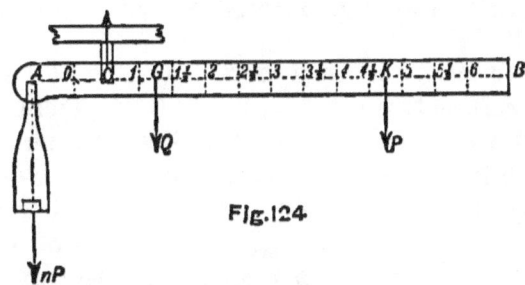

Fig. 124.

of weights is obviated. It consists of a straight beam AB, which is free to turn about a fulcrum C. The weight to be ascertained is placed in a scale-pan, which hangs from the end A. A fixed movable weight slides along the beam, which is graduated so that the graduation at which the movable weight is situate, when the beam rests in a horizontal position, gives the required weight.

To shew how the graduations are obtained.

Let P be the movable weight, Q the weight of the beam and scale-pan, G the point of the beam through which Q acts.

Let K be the position of the graduation n, i.e. the position P occupies when there is a weight nP in the scale-pan, and the beam balances in a horizontal position. Taking moments about C, we have

$$nP \cdot AC - Q \cdot CG - P \cdot CK = 0.$$

Putting $n = 0$, in this equation, we get the position O of the zero of the scale, $CO = -\dfrac{Q}{P} \cdot CG$, or O is on the other side of C to G, and at a distance $\dfrac{Q}{P} \cdot CG$ from it.

Hence $\qquad nP \cdot AC = P \cdot OK,$

or $\qquad\qquad OK = nAC.$

The graduations are obtained then by marking off distances from O, equal to AC, $2AC$, $3AC$, &c. By giving n fractional values we can obtain intermediate graduations.

152. *The Danish Steelyard.* This steelyard consists of a beam AB, terminating in a ball B; from the end A hangs the scale-pan in which the body to be weighed is placed. The fulcrum C is moved until the weight placed in the scale-pan is counterbalanced by that of the steelyard. The beam is graduated so that the position of C, when the beam balances, gives the corresponding weight in A.

To obtain the graduations.

Let P be the weight of the steelyard and scale-pan, acting through the point G of the steelyard. It is obvious that the zero graduation is at G, since the fulcrum must be at G, when the beam balances without any weight in the scale-pan.

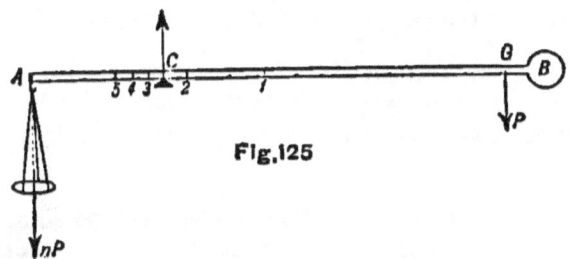

Fig. 125

Let C be the position of the graduation n, i.e. the point where the fulcrum is when there is a weight nP in the scale-pan, and the beam balances.

Taking moments about C, we have
$$nP \cdot AC = P \cdot CG = P(AG - AC),$$
$$\therefore AC = \frac{AG}{n+1}.$$

Hence the graduations are at a distance from A equal to
$$\frac{AG}{2}, \frac{AG}{3}, \frac{AG}{4}, \&c.$$
i.e. the distances between successive graduations are in Harmonical Progression.

Ex. 1. If the beam of a balance be horizontal, when there are no weights in the scale-pans, shew that if the balance be a false one, the actual weight of a body is the geometric mean of its apparent weights when weighed first in one scale-pan, and then in the other.

Ex. 2. If the arms of a false balance be without weight and one arm longer than the other by ¼th part of the shorter arm, and if in using it the substance to be weighed is put as often into one scale as the other, shew that the seller loses ⅝ per cent. on his transactions.

Ex. 3. If the bar of the common steelyard be 18 inches long, weigh 3 lbs. and be suspended at a point 3 inches from one extremity, what is the greatest weight which can be measured by a movable weight of 2 lbs.?

Ans. 16 lbs.

Ex. 4. A common steelyard is 12 inches long, and with the scale-pan weighs 1 lb., the centre of gravity of the two being 2 inches from the end to which the scale-pan is attached; find the position of the fulcrum when the movable weight is 1 lb. and the greatest weight that can be ascertained by means of the steelyard is 12 lbs. *Ans.* 1 in. from scale-pan.

Ex. 5. The movable weight of a common steelyard is 6 oz. A tradesman diminishes its weight by half an ounce: of how much is a person defrauded who buys what appears to weigh 6 lbs. by this steelyard?
Ans. $\frac{1}{2}$ oz.

Ex. 6. Find the length of a Danish steelyard, weighing 1 lb., when the distance between the graduations 4 lbs. and 5 lbs. is 1 inch.
Ans. 30 in.

153. Roberval's Balance. This consists of four uniform rods, AB, BD, DC, CA, freely jointed at their extremities and forming a parallelogram. The rods AB, CD can turn about pivots at their middle points E, F, which are fixed in a vertical support. The rods AB, CD are similar in every respect, as are the rods AC, BD. Equal scale-pans are rigidly connected with AC and BD.

The advantage of this balance is that it does not matter whereabouts the scale-pans the weights to be compared are placed.

Let the weight P, when placed in the scale-pan attached to AC, counterbalance the weight Q placed in the other scale-pan. If now the

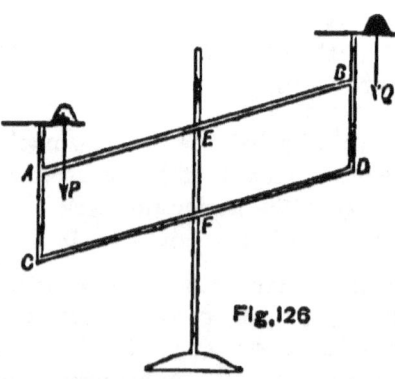

Fig. 126

system be supposed displaced by the beams AB, CD turning through a small angle, it is clear that the centres of mass of AB, CD suffer no displacement, while that of BD and its scale-pan is raised or lowered through a vertical distance p, say, and the centre of mass of AC and its

MACHINES. 261

scale-pan is lowered or raised through the same distance. The virtual work done by the weight of BD will be equal to, but of opposite sign to, that done by the weight of AC. Also the algebraical sum of the virtual work done by the internal forces of the system is zero. The equation of virtual work is therefore $Pp - Qp = 0$, since P, Q move through the same vertical distance as AC, and BD viz. p; therefore $P = Q$. This result holds wherever P and Q are placed in their respective scale-pans, i.e. whatever be their distances from the vertical support.

154. *The Differential Wheel and Axle.* In order to raise a very large weight by means of a comparatively small power, with the help of the ordinary 'wheel and axle', it would be necessary to make either the radius of the wheel inconveniently large, or else that of the axle so small

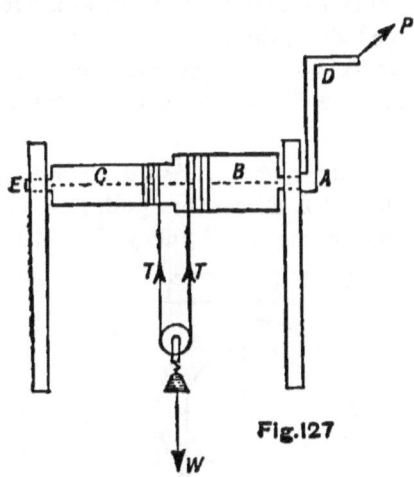

Fig.127

that it would be unable to bear the strain put upon it. This difficulty is got over in the 'Differential Wheel and Axle'. This consists of two axles B and C, of different radii, rigidly connected together and turning about their common axis AE, which is horizontal and turns in fixed sockets. The power P is applied at right angles to the axis, and at the end of an arm AD, the 'wheel'; the weight W is attached to a pulley supported by a rope which is wrapped one way round B, and the other way round C: P and the rope round the thicker axle B tend to turn the machine in opposite directions.

To find the Conditions of Equilibrium.

Let a be the length of AD, b, c the radii of B, C respectively, and T the tension of the rope supporting the pulley.

262 STATICS.

Since the pulley is in equilibrium
$$2T = W.$$
Since the machine is in equilibrium, taking moments about the axis AE, we have
$$Pa - Tb + Tc = 0,$$
$$\therefore Pa = T(b - c) = \tfrac{1}{2} W (b - c),$$
or
$$P : W = b - c : 2a.$$

Hence by making the radii of B and C as nearly equal as we please, the weight which a given power P can raise, may be increased to any extent.

The principle of work also enables us to obtain this result very easily.

In practice, this machine is useless, as in order to raise the weight through an appreciable height, the length of rope required would be very great. This difficulty is however got over in a modification of the differential wheel and axle, known from the name of the patentee, as *Weston's Differential Pulley*.

In the *Differential Pulley* shewn in figure 128, an endless chain passes over a fixed pulley B, under a movable pulley to which the weight is attached, and then over another fixed pulley C, a little smaller than, but

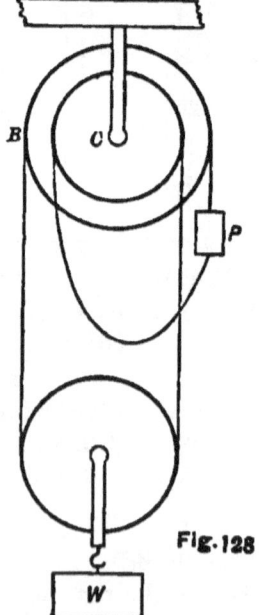

Fig. 128

coaxial with B: the two ends of the chain are jointed so as to form a loop, the Power is applied to the right-hand portion of the loop: to prevent the chain from slipping, there are cavities formed in the circumferences of the upper pulleys into which the links of the chain fit.

The condition of equilibrium is obtained as in the differential wheel and axle, and is the same, if we write b for a, i.e. is

$$P : W = b - c : 2b,$$

where b is the radius of the larger fixed pulley, c that of the smaller.

155. *Hunter's Differential Screw.* This consists of a screw AD which works in a fixed nut CC'. AD is hollow and has a thread cut inside it, in which a solid screw DE works. DE is prevented from turning round by some means, for instance, by means of a rod FEF' rigidly

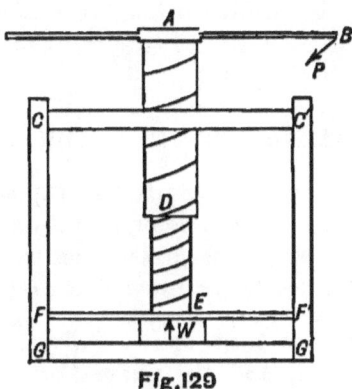

Fig. 129

connected with it, and whose ends work in smooth grooves, so that the screw DE can only move in a direction parallel to its axis.

The *weight* W is the resistance exerted by any substance placed between E and the base GG' of the framework $CGG'C'$. The *power* P is applied at the extremity of the arm AB which is at right angles to and rigidly connected with the screw AD.

Condition of Equilibrium. Let a be the length of AB, h, h' the distances between consecutive threads of AD, DE respectively.

Let us see the effect of the arm AB making a complete revolution. AD will clearly descend through a distance h: DE cannot turn with AD, and therefore will move upwards relatively to AD through a space h', i.e. will actually descend through a space $h - h'$: this is therefore the distance through which the weight is moved.

Let us suppose the virtual displacement made to be that which would be produced by P moving its point of application through a small angle θ, so that in consequence the weight descends through a distance x: as the distance through which DE descends is proportional to the angle through which AD turns, $x/(h-h')=\theta/2\pi$. As P and W are the only forces that do work during the above displacement, the equation of virtual work is

$$P \cdot a\theta - Wx = 0,$$
$$\therefore P \cdot 2\pi a = W(h-h').$$

This relation might have been obtained by an extension of the method adopted in Art. 146.

It is clear that by making h and h' sufficiently nearly equal, we can make W/P as great as we please; whereas the same result is obtained in the simple screw only by making a inconveniently large, or by making h so small that the thread is too weak to support the pressure on it.

EXAMPLES ON CHAPTER VII.

1. If a power P acting horizontally will support a weight W on a plane of inclination α, and would also support it on a plane of inclination β, acting parallel to the plane, the pressure on the plane in the former case being double that in the latter, prove that $\alpha = \frac{1}{2}\cos^{-1}(\frac{1}{4})$.

2. If in the first system there be two pulleys, the fixed ends only of the strings being parallel, and the power horizontal, prove that the mechanical advantage is $\sqrt{3}$.

3. In the first system the weights of the pulleys beginning with the highest are in A. P. and a power P supports a weight W; the pulleys are then reversed, the highest being placed lowest and so on, and now W and P when interchanged are in equilibrium: shew that $n(W+P)=2W'$, where W' is the total weight of the pulleys and n is the number of them.

4. If there be n pulleys in the third system, and if the string which goes over the lowest have the end at which the power is usually hung, passed under another movable pulley, over a fixed pulley, and then attached to the weight W; and if the weight of each pulley be w and no other power be used, prove that $W=(3 \cdot 2^{n-1}-n-1)w$.

5. In a weighing machine constructed on the principle of the common steelyard the pounds are read off by graduations reaching from

0 to 14, and the stones by weights hung at the end of the arm; if the weight corresponding to one stone be 7 oz., the movable weight $\frac{1}{2}$ lb., and the length of the arm one foot, prove that the distance between successive graduations is $\frac{3}{4}$ inches.

6. Shew that, in the third system, if there are n pullies, each of diameter $2a$ and weight w, the distance of the point of suspension of the weight from the line of action of the power is equal to

$$na\,\frac{2^{n+1}W+[(n-3)\,2^{n}+n+3]\,w}{2\,(2^{n}-1)\,W}.$$

7. In the first system of pulleys, shew that, if the weights of the pulleys are all equal, the equilibrium will not be affected by increasing P, W, and the weight of each pulley, by the same amount.

8. A weight W is weighed by a common steelyard, but a weight Q is substituted for the proper movable weight P. Shew that the error is $(W-wb/a)\,(P-Q)/Q$, where w is the weight of the steelyard, and b, a the distances from the fulcrum of the centre of gravity and of the scale-pan in which w is placed.

9. A false balance, the weight of whose beam may be neglected, has given weights in the pans, which weights are afterwards interchanged. In the two positions of equilibrium the beam makes complementary angles with the vertical. Shew that the line, joining the point of suspension to the middle point of the beam, makes with the beam twice the angle, that the beam makes with the vertical in one of its positions.

10. The weight of a common steelyard is Q, and the distance of its fulcrum from the point from which the weight hangs is a, when the instrument is in perfect adjustment; the fulcrum is displaced to a distance $a+\alpha$ from this end; shew that the correction to be applied to give the true weight of a body, which in the imperfect instrument appears to weigh W, is $(W+P+Q)\,\alpha/(a+\alpha)$, P being the movable weight.

11. If in the first system, P is the power (acting upwards), W the weight, and R the stress on the beam from which the pulleys hang, shew that R is greater than $W\,(1-2^{-n})$ and less than $(2^n-1)\,P$.

12. If on a steelyard the movable weight P which forms the power be increased in the ratio $1+k:1$, prove that the consequent error in W, the weight to be found, is kY, where Y is the weight that must be removed from W in order to preserve equilibrium when P is moved close to the fulcrum.

266   STATICS.

13. Prove that, if in a machine the weight can be supported by the friction alone, then in raising the weight half the power at least is wasted in overcoming friction.

Apply this to the differential pulley; and prove that if the weight can be supported by friction alone, the radius of the axle must be greater than the difference of the radii of the pulleys multiplied by the cosecant of the angle of friction.

14. A single movable pulley, weight W, is just supported by the power P, which is applied at one end of a cord which goes under the pulley and is then fastened to a fixed point: shew that if ϕ be the angle subtended at the centre by the part of the string in contact with the pulley, it is given by the equation

$$P(1 - 2e^{\mu\phi}\sin\phi + e^{2\mu\phi})^{\frac{1}{2}} = W.$$

15. A true balance is in equilibrium with unequal weights P, Q in its scales. If a small weight be added to P, the consequent vertical displacement of Q is equal to that which would be the vertical displacement of P, were the same small weight to be added to Q instead of to P.

16. Prove that in the third system, if the pulleys be small compared with the lengths of the strings, the necessary correction for the weights of the strings is the addition to $W, P_2, P_3, \ldots P_n$ respectively, of the weights of lengths $h_2 + h_3 \ldots + h_n + h$, $2(h_2 - h_3)$, $2(h_3 - h_4)$, $\ldots 2(h_n - h)$ of string: where $h_1, h_2, h_3, \ldots h_n$ are the heights of the n pulleys (whose weights are $P_1, P_2, \ldots P_n$ respectively) above the line of attachment, supposed horizontal, of the strings to the weight W, and h the height of the point of attachment of the power above the same line.

17. In graduating a steelyard to weigh pounds marks are made with a file, a weight x being removed for each notch. With the movable weight P at the end of the beam, n lbs. can be weighed after the graduation is completed, $(n+1)$ before it is begun, shew that $n(n+1) = 2P/x$, and find the error made in weighing m pounds. The c. g. of the steelyard is originally under the point of suspension.

18. A Danish steelyard, weight W lbs. and accurately graduated, is painted. In consequence of the paint, the apparent weights of two known weights of X lbs. and Y lbs. are found when weighed by the steelyard to be $(X - x)$ lbs., $(Y - y)$ lbs. respectively. Prove that the centre of gravity of the paint divides the graduated arm in the ratio $W(x - y) : Yx - Xy$; and that its weight is, to a first approximation, $x(W + Y)/(X - Y) + y(W + X)/(Y - X)$.

19. A brass figure $ABDC$, of uniform thickness, bounded by a circular arc BDC (greater than a semicircle) and two tangents AB, AC inclined at an angle $2a$, is used as a letter-weigher as follows. O the centre of the circle is a fixed point, about which the machine can turn freely, and a weight P is attached to the point A, the weight of the machine itself being w. The letter to be weighed is suspended from a clasp (whose weight may be neglected) at D on the rim of the circle, OD being perpendicular to OA. The circle is graduated and is read by a pointer which hangs vertically from O: when there is no letter attached, the point A is vertically below O, and the pointer indicates zero. Obtain a formula for the graduation of the circle, and shew that if $P = \frac{1}{3} w \sin^2 a$, the reading of the machine will be $\frac{1}{4} w$ when OA makes with the vertical an angle equal to

$$\tan^{-1}\left\{\frac{(\pi + 2a)\sin^3 a + 2 \sin a \cos a}{(\pi + 2a)\sin^3 a + 2 \cos a}\right\}.$$

20. A common steelyard is graduated on the assumptions that its weight is Q, and that the movable weight is W, both of which assumptions are incorrect. If two masses whose real weights are P and R appear to weigh $P+X$, and $R+Y$, then the weight of the steelyard and the movable weight are less than their assumed values by

$$\frac{W}{D}(X-Y) \text{ and } \frac{Q}{D}(X-Y) + \frac{a}{bD}(PY - RX)$$

respectively, where b, a are the distances of the fulcrum from the centre of gravity of the bar and the point of attachment of the substance to be weighed, and $D = P - R + X - Y$.

APPENDIX.

1. IF a solid cube of finite size be cut by parallel planes into n slices of equal thickness, we can by sufficiently increasing n make the volume of each slice smaller than any assignable volume. The volume of a slice is in this case said to be ultimately an *indefinitely small quantity*.

An indefinitely small quantity, then, is one which though itself less than any assignable quantity, yet when multiplied by a sufficiently great number amounts to a finite quantity. It is often said to be *ultimately zero*, but it must be understood that it is not *absolute zero*, which does not amount to a finite quantity, by however great a quantity it is multiplied.

Let the above cube be now cut by planes parallel to another face, so that each slice is divided into n equal prisms, each having square ends. Again, let the cube be cut by planes parallel to a third face, so that each prism is divided into n equal cubes. The total number of cubes is n^3, of prisms n^2, and of slices n; and it requires n prisms to make a slice, or n^2 cubes. It follows then that, when n is increased indefinitely, a slice, a prism, and a cube become all indefinitely small, but that though n slices make up a finite volume, n prisms do not, and though the sum of n^2 prisms is finite, that of n^2 cubes is indefinitely small. Therefore the ratio of a prism to a slice is indefinitely small, and also that of a cube to a prism, and *à fortiori* that of a cube to a slice. This is usually expressed by saying that a slice, a prism, and a cube are respectively of the *first, second* and *third orders* of indefinitely small quantities.

One indefinitely small quantity is *of a higher order* than another, when the ratio of the first to the second is indefinitely small.

Two quantities are equal when their difference is indefinitely small compared with either: i.e. two finite quantities are equal when their difference is an indefinitely small quantity, and two indefinitely small quantities are equal when their difference is a small quantity of higher order.

When we assert that the algebraical sum of a finite number of indefinitely small quantities is zero, we are not stating a truism, but mean that they are so related that their algebraical sum is of a higher order than that of the quantities involved.

2. Prop. *If two series, consisting of the same number of indefinitely small quantities of the same order, are such, that each term of the one bears to the corresponding term of the other a ratio differing from k (a finite quantity) by an indefinitely small quantity, the sum of the one series is k into the sum of the other.*

Let $a_1, a_2, \ldots a_n$ be the first series, $b_1, b_2, \ldots b_n$ the second, so that

$$\frac{a_1}{b_1} = k + c_1, \quad \frac{a_2}{b_2} = k + c_2, \ldots \ldots \frac{a_n}{b_n} = k + c_n,$$

where $c_1, c_2 \ldots c_n$ are indefinitely small: let c be the greatest of these quantities.

Then

$$a_1 + a_2 + \ldots a_n = k(b_1 + b_2 + \ldots b_n) + b_1 c_1 + b_2 c_2 + \ldots b_n c_n;$$

$$\therefore \ \Sigma(a) - k \Sigma(b) \text{ is not } > c \Sigma(b);$$

$$\therefore \ \Sigma(a) = k \Sigma(b),$$

since $c\Sigma(b)$ is indefinitely small compared with $k\Sigma(b)$.

Cor. Hence two infinite series of indefinitely small quantities of the first order, such that each term of the one differs from the corresponding term of the other by a quantity of the second order, are equal.

This explains why in Arts. 97, 99, 101 and 102 we have neglected one infinite series and retained another: this is done when the first series is of a higher order than the second.

3. As an illustration of these principles we will give proofs of Guldin's theorems.

One theorem is, that *the volume, generated by the complete revolution of a plane area about any straight line in its plane and not cutting it, is equal to that of a right cylinder whose section is the plane area and height the length of the path described by the centre of mass of the area.*

Draw a number of straight lines at right angles to the line AB, about which the revolution takes place, dividing the area S into n strips of equal breadth. Let Pp, Qq be two consecutive lines of this system, typical of the rest, M, N the points where they meet AB.

Draw PR, pr perpendicular to Qq.

The volume, generated by the revolution of $PRrp$ about AB, differs from that generated by $PQqp$ by the volumes generated by the two curvilinear triangles PQR, pqr.

But when n is increased indefinitely, the breadth only of the rectangle is diminished indefinitely, whereas both length and breadth of each triangle is diminished indefinitely; the volumes generated by the latter are therefore of a higher order than that generated by the former.

Hence the total volume generated by the area equals

the sum of the volumes generated by the rectangles of which Pr is a type, i.e.

$$= \Sigma \left(\pi PM^2 . MN - \pi pM^2 . MN \right)$$
$$= \pi \Sigma \left\{ (PM - pM)(PM + pM) MN \right\}$$
$$= \pi \Sigma \left\{ Pp . MN (PM + pM) \right\}.$$

Also the sum of the areas of the rectangles is the area of the figure S; and since they differ by the sum of the areas of the triangles PQR, pqr, &c., which are of a higher order than the rectangles, the centres of mass of the sum of the rectangles and the figure S must be coincident.

Therefore x, the distance of the C.M. of S from AB

$$= \frac{\Sigma \left\{ Pp . MN . \tfrac{1}{2}(PM + pM) \right\}}{\Sigma (Pp . MN)}$$

$$= \frac{\tfrac{1}{2} \text{ vol. generated by } S}{\pi \times \text{area } S},$$

$\therefore$ volume generated by $S = S . 2\pi x.$

The second theorem is, that *the area of the surface, generated by the revolution of a curve about any straight line in its plane and not cutting the curve, is equal to the rectangle, whose length is the length of the curve and breadth the distance of the curve's centre of mass from the straight line.*

Let PQ be a side of a polygon, either inscribed within or circumscribed about the curve: let R be the middle point of PQ, and therefore its centre of mass. Draw RK perpendicular to the line AB, about which the curve revolves.

As in Art. 99, it can be shewn that the area of the surface generated by the revolution of PQ about AB is $2\pi PQ . RK.$

Therefore the total surface generated by the revolution of the polygon about AB

$$= \Sigma\,(2\pi PQ\,.\,RK)$$
$$= 2\pi\Sigma\,(PQ\,.\,RK)$$
$$= 2\pi x \times \text{perimeter of polygon,}$$

where x is the distance of its centre of mass from AB.

When the lengths of the sides of the inscribed and circumscribed polygons are diminished indefinitely and their number increased indefinitely, their perimeters differ by indefinitely small quantities, and their centres of mass become coincident. The surfaces generated by each are therefore equal.

It is assumed as axiomatic, that as the perimeter of the curve lies in position between the two polygons which ultimately coincide, it is equal to the perimeter of either polygon, its centre of mass coincides with that of either polygon, and the surface generated by it is equal to that generated by either polygon.

Hence the surface generated by the curve is equal to the product of the length of the curve into the length of the path traced out by its centre of mass.

Each of Guldin's theorems can easily be extended to the case in which the revolution is not a complete one. There is no limitation in either as to the number of times, in which a straight line at right angles to AB cuts the generating curve.

Ex. Find the volume and surface of an anchor-ring, the figure generated by the revolution of a circle about a line in its plane, and not intersecting it.

Ans. Vol.$=2\pi^2 a^2 c$, surface$=4\pi^2 ac$, where a is the radius of the circle, and c the distance of its centre from the line.

4. The following proposition has been assumed throughout. *The limit of*
$$\frac{1^p + 2^p + 3^p + \ldots (n-1)^p}{n^{p+1}} = \frac{1}{p+1};$$

APPENDIX.

where p is any positive quantity, and n is increased indefinitely.

Let S_p denote $1^p + 2^p + 3^p + \ldots (n-1)^p$.

$$n^{p+1} - (n-1)^{p+1} = (p+1)(n-1)^p + \frac{(p+1)p}{2!}(n-1)^{p-1} + \&c.$$

$$(n-1)^{p+1} - (n-2)^{p+1} = (p+1)(n-2)^p$$
$$+ \frac{(p+1)p}{2!}(n-2)^{p-1} + \&c.$$

$$\ldots\ldots\ldots\ldots = \ldots\ldots\ldots\ldots\ldots\ldots\ldots\ldots$$

$$2^{p+1} - 1^{p+1} = (p+1) 1^p + \frac{(p+1)p}{2!} 1^{p-1} + \&c.$$

$\therefore$ by addition

$$n^{p+1} - 1^{p+1} = (p+1) S_p$$
$$+ \frac{(p+1)p}{2!} S_{p-1} + \frac{(p+1)p(p-1)}{3!} S_{p-2} + \&c.,$$

$$\therefore \frac{1}{p+1} = \frac{1}{(p+1)n^{p+1}} + \frac{S_p}{n^{p+1}}$$
$$+ \frac{p}{2!} \cdot \frac{1}{n} \cdot \frac{S_{p-1}}{n^p} + \frac{p(p-1)}{3!} \cdot \frac{1}{n^2} \cdot \frac{S_{p-2}}{n^{p-1}} + \&c.,$$

But $\dfrac{S^p}{n^{p+1}}$ is obviously $< \dfrac{(n-1)^{p+1}}{n^{p+1}}$, i.e. is < 1.

Similarly $\dfrac{S^{p-r}}{n^{p-r+1}}$ is < 1, if p is $> r$.

Hence, when p is integral,

$$\frac{1}{p+1} = \frac{S_n}{n^{p+1}} + \frac{A_1}{n} + \frac{A_2}{n^2} + \frac{A_3}{n^3} + \ldots \frac{A_{p+1}}{n^{p+1}},$$

where A_1, A_2, A_3, &c. are all finite quantities;

$\therefore \dfrac{S_p}{n^{p+1}} = \dfrac{1}{p+1}$, when n is increased indefinitely.

If p be fractional

$$\frac{S_{p-r}}{n^{p+1}} = 0 \text{ ultimately, when } p \text{ is } > r,$$

when p is $< r$, $\frac{S_{p-r}}{n^{p+1}}$ is $< \frac{1}{n^p}$;

$$\therefore \frac{1}{p+1} - \frac{S_p}{n^{p+1}} \text{ is numerically } < \frac{1}{n^p}\left\{\frac{p}{2!} + \frac{p(p-1)}{3!} + \&c.\right\};$$

i.e. $< \dfrac{1}{(p+1)\,n^p}\left\{(1+1)^{p+1} - 1 - \dfrac{(p+1)}{1!}\right\};$

i.e. $= 0$ ultimately.

Hence the result holds, whether p is integral or not.

www.ingramcontent.com/pod-product-compliance
Lightning Source LLC
Chambersburg PA
CBHW032102220426
43664CB00008B/1106